Beyond Fear

Thinking Sensibly About Security
in an Uncertain World

Bruce Schneier

SPRINGER SCIENCE+BUSINESS MEDIA, LLC

© 2003 Springer Science+Business Media New York
Originally published by Springer-Verlag New York, Inc. in 2003
Softcover reprint of the hardcover 1st edition 2003

Library of Congress Cataloging-in-Publication Data

Schneier, Bruce
    Beyond fear : thinking sensibly about security in an uncertain
    world / Bruce Schneier.
        p.   cm.
    Includes index.
    ISBN 978-1-4757-8119-9   ISBN 978-0-387-21712-3 (eBook)
    DOI 10.1007/978-0-387-21712-3
    1. Terrorism--United States--Prevention.   2. War on Terrorism, 2001–
    I.   Title.
    HV6432.S36   2003
    363.3'2--dc21

                                                    2003051488

Manufactured in the United States of America.
Printed on acid-free paper.

9 8 7 6 5 4 3

ISBN 978-1-4757-8119-9

To My Parents

Contents

Part One

Sensible Security

Chapter 1

All Security Involves Trade-offs

In the wake of 9/11, many of us want to reinvent our ideas about security. But we don't need to learn something completely new; we need to learn to be smarter, more skeptical, and more skilled about what we already know. Critical to any security decision is the notion of trade-offs, meaning the costs—in terms of money, convenience, comfort, freedoms, and so on—that inevitably attach themselves to any security system. People make security trade-offs naturally, choosing more or less security as situations change. This book uses a five-step process to demystify the choices and make the trade-offs explicit. A better understanding of trade-offs leads to a better understanding of security, and consequently to more sensible security decisions.

The attacks were amazing. If you can set aside your revulsion and horror—and I would argue that it's useful, even important, to set them aside for a moment—you can more clearly grasp what the terrorists accomplished.

The attacks were amazing in their efficiency. The terrorists turned four commercial airplanes into flying bombs, killed some 3,000 people, destroyed $40 billion in property, and did untold economic damage. They altered the New York skyline as well as the political landscape of the U.S. and the whole world. And all this was done with no more than a thirty-person, two-year, half-million-dollar operation.

The attacks were amazing in the audacity of their conception. No one had ever done this before: hijack fuel-laden airplanes and fly them into skyscrapers. We'll probably never know for sure if the terrorists counted on the heat from the ensuing fire to fatally weaken the steel supports and bring down the World Trade Center towers, but those who planned the attacks certainly chose long-distance flights as targets, since they would be carrying heavy fuel loads. The scheme's audacity meant no one had planned a defense against this type of attack.

The attacks were amazing for their detailed planning and preparation and the discipline shown by the multilayered, compartmentalized organization that carried them out. The plan probably involved a total of some thirty people, and, of these, some had to have been willing to die. Others most likely had to be kept from knowing they were going to die. The keeping of secrets and careful parceling out of information doubtless required training. It required coordination. It required extraordinary discipline. Indeed, the sheer scope of the attacks seemed beyond the capability of a terrorist organization and in fact has forced us to revise our notions of what terrorist organizations are capable of.

At the same time, the entire operation was amazing in its technological simplicity. It required no advanced technology that couldn't be hijacked or (as in the case of GPS devices) easily purchased. All technical training could be easily had. And there was no need for complex logistical support: Once the attacks were set in motion, the terrorists were on their own; and once they were in the air, each group of four or five was on its own, independent and self-sufficient.

The attacks were amazing because they rewrote the hijacking rulebook. Previous responses to hijackings centered around one premise: Get the plane on the ground so negotiations can begin. The threat of airplane bombings, we had come to believe, was solved by making sure passengers were on the same flights as their baggage. These attacks made all that obsolete.

The attacks were amazing because they rewrote the terrorism book, too. Al Qaeda recruited a new type of attacker. Not the traditional candidate—young, single, fanatical, and with nothing to lose—but people older and more experienced, with marketable job skills. They lived in the West, watching television, eating fast food, drinking in bars. Some vacationed in Las Vegas. One left a wife and four children. It was also a new page in the terrorism book in other ways. One of the most difficult things about a terrorist operation is getting away at the end. This suicide attack neatly solved that problem. The U.S. spends billions of dollars on remote-controlled precision-guided munitions, while all Al Qaeda had to do was recruit fanatics willing to fly planes into skyscrapers.

Finally, the attacks were amazing in their success rate. They weren't perfect; 100 percent of the attempted hijackings were successful, but only 75 percent of the hijacked planes successfully reached their targets. We don't know if other planned hijackings were aborted for one reason or another, but that success rate was more than enough to leave the world shocked, stunned, and more than a little bit fearful.

• • • •

The 9/11 terrorist operation was small, efficient, relatively low-tech, very strictly disciplined, highly compartmentalized, and extremely innovative. Did we stand a chance?

The plan's size, discipline, and compartmentalization were critical in preventing the most common failure of such an operation: The plan wasn't leaked. Al Qaeda had people in the U.S., in some cases for years, then in staged arrivals for months and then weeks as the team grew to full size. And, throughout, they managed to keep the plan secret. No one successfully defected. And no one slipped up and gave the plan away.

Not that there weren't hints. Zacarias Moussaoui, the "twentieth hijacker," was arrested by the FBI in Minnesota a month before the attacks. The local FBI office wanted to investigate his actions further. German intelligence had been watching some parts of the operation, and U.S. and French intelligence had been watching others. But no one "connected the dots" until it was too late, mostly because there really were no dots to connect. The plan was simply too innovative. There was no easy-to-compare template and no clear precedent, because these terrorists in a very real sense wrote the book—a new book.

Rarely does an attack change the world's conception of attack. And yet while no single preparation these terrorists made was in and of itself illegal, or so outlandish that it was likely to draw attention—taken together, put together in just this way, it was devastating. Nothing they did was novel—Tom Clancy wrote about terrorists crashing an airplane into the U.S. Capitol in 1996, and the Algerian GIA terrorist group tried to hijack a plane and crash it into the Eiffel Tower two years before that—yet the attack seemed completely new and certainly was wholly unexpected. So, not only did our conception of attack have to change; in response, so did our conception of defense.

••••

Since 9/11, we've grown accustomed to ID checks when we visit government and corporate buildings. We've stood in long security lines at airports and had ourselves and our baggage searched. In February 2003, we were told to buy duct tape when the U.S. color-coded threat level was raised to Orange. Arrests have been made; foreigners have been deported. Unfortunately, most of these changes have not made us more secure. Many of them may actually have made us *less* secure.

The problem is that security's effectiveness can be extremely hard to measure. Most of the time, we hear about security only when it fails. We don't know how many, if any, additional terrorist attacks were prevented or aborted or scared off prior to 9/11. We don't know what, if anything, we could have done to foil the 9/11 attacks, and what additional security would have merely been bypassed by minor alterations in plans. If the 9/11 attacks had failed, we wouldn't know whether it had been because of diligent security or because of some unrelated reason. We might not have known about them at all. Security, when it is working, is often invisible not only to those being protected, but to those who plan, implement, and monitor security systems.

But it gets even more complicated than that. Suppose security is perfect, and there are no terrorist attacks; we might conclude that the security expenditures are wasteful, because the successes remain invisible. Similarly, security might fail without us knowing about it, or might succeed against the attacks we know about but fail in the face of an unforeseen threat. A security measure might reduce the likelihood of a rare terrorist attack, but could also result in far greater losses from common criminals. What's the actual risk of a repeat of 9/11? What's the risk of a different but equally horrific sequel? We don't know.

In security, things are rarely as they seem. Perfectly well-intentioned people often advocate ineffective, and sometimes downright countereffective, security measures. I want to change that; I want to explain how security works.

Security is my career. For most of my life, I have been a professional thinker about security. I started out focusing on the mathematics of security—cryptography—and then computer and network security; but more and more, what I do now focuses on the security that surrounds our everyday lives. I've worked for the Department of Defense, implementing security solutions for military installations. I've consulted for major financial institutions, governments, and computer companies. And I founded a company that provides security monitoring services for corporate and government computer networks.

Since the attacks of 9/11, I have been asked more and more about our society's security against terrorism, and about the security of our society in general. In this book, I have applied the methods that I and others have developed for computer security to security in the real world. The concepts, ideas, and rules of security as they apply to computers are essentially no different from the security concepts, ideas, and rules that apply, or should apply, to the world at large. The way I see it, security is all of a piece. This attitude puts me, I suspect, in a minority among security professionals. But it is an attitude, I believe, that helps me to see more clearly, to reason more dispassionately than other security professionals, and to sort out effective and ineffective security measures.

This book is about security: how it works and how to think about it. It's not about whether a particular security measure works, but about how to analyze and evaluate security measures. For better or worse, we live in a time when we're very likely to be presented with all kinds of security options. If there is one result I would like to see from this book, it is that readers come away from reading it with a better sense of the ideas and the security concepts that make systems work— and in many cases not work. These security concepts remain unchanged whether you're a homeowner trying to protect your possessions against a burglar, the President trying to protect our nation against terrorism, or a rabbit trying to protect itself from being eaten. The attackers, defenders, strategies, and tactics are different from one security situation to another, but the fundamental principles and practices—as well as the basic and all-important ways to *think* about security—are identical from one security system to another.

Whether your concern is personal security in the face of increasing crime, computer security for yourself or your business, or security against terrorism, security issues affect us more and more in our daily lives, and we should all make an effort to understand them better. We need to stop accepting uncritically what politicians and pundits are telling us. We need to move beyond fear and start making sensible security trade-offs.

•••

And "trade-off" really is the right word. Every one of us, every day of our lives, makes security trade-offs. Even when we're not thinking of threats or dangers or attacks, we live almost our entire lives making judgments about security, assessments of security, assumptions regarding security, and choices about security.

When we brush our teeth in the morning, we're making a security trade-off: the time spent brushing in exchange for a small amount of security against tooth decay. When we lock the door to our home, we're making a security trade-off: the inconvenience of carrying and using a key in exchange for some security against burglary (and worse). One of the considerations that goes into which car we purchase is security against accidents. When we reach down at a checkout counter to buy a candy bar and notice that the package has been opened, why do we reach for another? It's because a fully wrapped candy bar is a better security trade-off, for the same money, than a partially wrapped one.

Security is a factor when we decide where to invest our money and which school to send our children to. Cell phone companies advertise security as one of the features of their systems. When we choose a neighborhood to live in, a place to vacation, and where we park when we go shopping, one of our considerations is security.

We constantly make security trade-offs, whether we want to or not, and whether we're aware of them or not. Many would have you believe that security is complicated, and should be best left to the experts. They're wrong. Making security trade-offs isn't some mystical art like quantum mechanics. It's not rocket science. You don't need an advanced degree to do it. Everyone does it every day; making security trade-offs is fundamental to being alive. Security is pervasive. It's second nature, consciously and unconsciously part of the myriad decisions we make throughout the day.

The goal of this book is to *demystify* security, to help you move beyond fear, and give you the tools to start making sensible security trade-offs. When you're living in fear, it's easy to let others make security decisions for you. You might passively accept any security offered to you. This isn't because you're somehow incapable of making security trade-offs, but because you don't understand the rules of the game. When it comes to security, fear is the barrier between ignorance and understanding. To get beyond fear, you have to start thinking intelligently about the trade-offs you make. You have to start evaluating the risks you face, and the security options you have for dealing with those risks. There's a lot of lousy security available for purchase and a lot of lousy security being imposed on us. Once you move beyond fear and start thinking sensibly about trade-offs, you will be able to recognize bad or overpriced security when you see it. You will also be able to spot ineffectual security—and explain why it's ineffectual.

As citizens, sometimes we have choices in our security trade-offs and sometimes we don't. Much security is imposed on us by law or business practice, and even if we dissent we cannot choose not to comply with the trade-offs. We cannot opt out of the FBI database, or decide that we would rather not have ourselves recorded on the thousands of security cameras in any major city. Still, there are limited trade-offs we *can* make. Banking security is what it is; we can't choose more or less of it. But we can choose to place our money elsewhere. Airline security is largely dictated by government regulations; airlines don't compete with each other based on security. But we can decide to drive instead. Many ways to defend our homes may be illegal, but lawful options include various brands of door locks or wall safes. And, as citizens, we can influence social and government policies.

I'm going to help you make sensible security trade-offs by teaching you how to think about security. Much of what I'm going to say is in stark contrast to the things you hear on television and the things our politicians are telling us. I read the newspapers, too, and the things we're asked to accept continually incense me. I've watched the creation of a new cabinet-level Department of Homeland Security. I've seen budget expenditures for security reaching $33.7 billion in 2003, and even more in future years. We're being asked to pay a lot for security, and not just in dollars. I'd like to see us get our money's worth.

Security is both a feeling and a reality. We're secure when we feel protected from harm, free from dangers, and safe from attack. In this way, security is merely a state of mind. But there's the reality of security as well, a reality that has nothing do with how we feel. We're secure when we actually *are* protected. Although both are important, this book is more about the reality of security than the feeling. We need to feel in control and positive and not harried and fearful for security to have much of a positive effect on our daily lives; living in a state of constant fear, after all, would take away many of the benefits we want from real security. But it's nonetheless important to ground that feeling of security in the reality of security, and not merely in placebos.

In some ways, this is analogous to health. If you went to the doctor because you had a badly damaged leg, she wouldn't pretend that she could return your leg to its undamaged state if she couldn't. She would tell you the truth, describe your treatment options, and help you choose the one that's best for you. Ignoring reality is not an effective way to get healthier, or smarter, or safer, even though it might temporarily make you feel better.

• • • •

Security is not an isolated good, but just one component of a complicated transaction. It costs money, but it can also cost in intangibles: time, convenience, flexibility, or privacy. You will be able to intelligently assess these trade-offs. I won't tell you what you *want* to buy or which national security policies to support—these are personal decisions—but I will give you the tools you need to make those decisions for yourself. No security is foolproof, but neither is all security equal. There's cheap security and expensive security. There's unobtrusive security and security that forces us to change how we live. There's security that respects our liberties and there's security that doesn't. There's security that really makes us safer and security that only lets us feel safer, with no reality behind it.

You face the same challenges with other choices. Your doctor can urge you to quit smoking, to exercise regularly, or to go on a low-cholesterol diet. But each of these actions requires trade-offs, and as a patient you may not choose to do the things your doctor recommends. You may not want to put in the time. You may be unwilling to make the trade-offs necessary to change your habits. You may not be able to afford the cost of treatment. You may be seduced by a guy selling weight-loss systems or herbal remedies on late-night TV.

A common path to bad security is knee-jerk reactions to the news of the day. Too much of the U.S. government's response post-9/11 is exactly that. We are told that we are in graver danger than ever, and that we must change our lives in drastic and inconvenient ways in order to be secure. We are told that we must sacrifice privacy and anonymity and accept restrictions on our actions. We are told that the police need new far-reaching investigative powers, that domestic spying capabilities need to be instituted, that our militaries must be brought to bear on countries that support terrorism, and that we must spy on each other. The security "doctors" are telling us to trust them, that all these changes are for our own good.

But the reality is that most of the changes we're being asked to endure won't result in good security. They're Band-Aids that ignore the real problems. Some of these changes may enhance our feeling of security, but would actually make us less safe. Our legislatures are not spending enough time examining the consequences of these so-called security measures. As you'll see in the next chapter, security is always a trade-off, and to ignore or deny those trade-offs is to risk losing basic freedoms and ways of life we now take for granted.

When you understand security and are able to make sensible trade-offs, you become actively involved in these security decisions and won't accept uncritically whatever is being offered to you. You may not be able to directly affect public policy, but you can decide whether to use the security provided or to do something else. You will be able to intelligently make your own trade-offs. And en masse, you can eventually change public policy.

. . . .

The world is a dangerous place, but it's also a good and decent place. People are good. People are kind. People are nice. Even in the worst neighborhoods, most people are safe. If Al Qaeda has 5,000 members, this is still only one in a million people worldwide, and if they were all in America, only one in 60,000. It's hard to find a terrorist, kidnapper, or bank robber, because there simply aren't that many in our society.

The point here is—and it's an important one to keep in mind—security exists to deal with the few bad apples, but if we allow those antisocial types to dictate social policy for everyone, we're making a mistake. Perfect security is impractical because the costs are simply too high; we would have to treat the whole world as a threatening place and all the people in it as evildoers, when in fact the real threats are not nearly so pervasive. We'd have to create an extremely oppressive regime. But freedom is security. Openness is security. If you want proof, look around you. The world's liberal democracies are the safest societies on the planet. Countries like the former Soviet Union, the former East Germany, Iraq, North Korea, and China tried to implement large-scale security systems across their entire populaces. Would anyone willingly trade the dangerous openness of the U.S. or most countries in Europe for the security of a police state or totalitarian government?

When you're scared, it's easy to lock the doors and head for the cellar. It's easy to demand that "something must be done" and that "those responsible must pay." It's easy to reach for the feeling of security and ignore its reality. But, remember, security catastrophes are extremely rare events. Making large changes in response to rare events often doesn't make sense, because these infrequent occurrences seldom have anything to do with day-to-day life. People shouldn't let their fears dominate their everyday life; there just aren't enough bad apples around.

Security is complicated and challenging. Easy answers won't help, because the easy answers are invariably the wrong ones. What you need is a new way to think.

••••

All security is, in some way, about prevention. But prevention of what, exactly? Security is about *preventing adverse consequences from the intentional and unwarranted actions of others*. What this definition basically means is that we want people to behave in a certain way—to pay for items at a store before walking out with them, to honor contracts they sign, to not shoot or bomb each other—and security is a way of ensuring that they do so. Obviously, if people always did what they were supposed to do, there would be no need for security. But because not everyone does, security is critical to every aspect of society. The definition above tries to brings these issues into focus.

- Security is about prevention. A *security system* is the set of things put in place, or done, to prevent adverse consequences. Cars have antitheft security systems. The money in your wallet has a variety of anti-counterfeiting security systems. Airports and airplanes have many overlapping security systems. Some security systems, like financial audits, don't stop criminals but reduce or prevent the adverse consequences of their crimes. (This type of after-the-fact security system has a deterrence effect, which acts as a form of prevention.) Like any other system, security systems can be attacked, can have flaws, and can fail.
- Security concerns itself with *intentional* actions. This points to an important distinction: Protecting assets from unintentional actions is safety, not security. While it is common to talk about securing ourselves against accidents or from natural disasters, in this book I try to restrict the discussion to security from intentional actions. In some ways this is an arbitrary distinction, because safety and security are similar, and the things that protect from one also protect from the other. And, in fact, safety is often more important than security, since there are far more unintentional unwarranted actions in the world than intentional ones. I make the distinction for the purpose of limiting the scope of this book.
- These intentional actions are *unwarranted* from the point of view of the *defender*. Note that the unwarranted actions are not necessarily illegal.
- Security requires the concept of an *attacker* who performs these intentional and unwarranted actions. It's important to understand that this term is meant to be nonpejorative, value-neutral. Attackers may be on your side. Burglars are criminals, but police officers who attack a drug lord in his mountain hideout are the good guys. Sometimes the attacker isn't even human, as in the case of a tiger trying to pry open your Land Rover and eat you. The term *attacker* assumes nothing about the morality or the legality of the attack. It doesn't even imply malicious intent, although that certainly is often involved.

- Those intentional unwarranted actions are called *attacks*. An attack is a specific way to attempt to break the security of a system or a component of a system. It can refer to an abstraction—"You can successfully attack a house by breaking a window"—or it can refer to a specific incident—"On 7 September 1876, the Jesse James Gang attacked the First National Bank of Northfield in Northfield, Minnesota."
- The objects of attack are *assets*. Assets can be as small as a single diamond and as large as a nation's infrastructure.
- Security can also refer to the mechanisms used to provide the protection. In this book, the individual, discrete, and independent security components are called *countermeasures*. Countermeasures include door locks, guards, ID cards, and the metallic thread in banknotes that makes them more difficult to counterfeit. Walls, alarms, cryptography, pots of boiling oil you can pour on the heads of people invading your castle—these are all countermeasures. A security system consists of a series of countermeasures.

Unfortunately, many countermeasures are ineffective. Either they do not prevent adverse consequences from the intentional and unwarranted actions of people, or the trade-offs simply aren't worth it. Those who design and implement bad security don't seem to understand how security works or how to make security trade-offs. They spend too much money on the wrong kinds of security. They make the same mistakes over and over. And they're constantly surprised when things don't work out as they'd intended.

One problem is caused by an unwillingness on the part of the engineers, law enforcement agencies, and civic leaders involved in security to face the realities of security. They're unwilling to tell the public to accept that there are no easy answers and no free rides. They have to be seen as doing something, and often they are seduced by the promise of technology. They believe that because technology can solve a multitude of problems and improve our lives in countless ways, it can solve security problems in a similar manner. But it's not the same thing. Technology is generally an enabler, allowing people to do things. Security is the opposite: It tries to prevent something from happening, or prevent people from doing something, in the face of someone actively trying to defeat it. That's why technology doesn't work in security the way it does elsewhere, and why an overreliance on technology often leads to bad security, or even to the opposite of security.

The sad truth is that bad security can be worse than no security; that is, by trying and failing to make ourselves more secure, we make ourselves less secure. We spend time, money, and energy creating systems that can themselves be attacked easily and, in some cases, that don't even address the real threats. We make poor trade-offs, giving up much in exchange for very little security. We surround ourselves with security countermeasures that give us a feeling of security rather than the reality of security. We deceive ourselves by believing in security that doesn't work.

....

Security is complex, but complex things can be broken down into smaller and simpler steps. Throughout this book, I use a five-step process to analyze and evaluate security systems, technologies, and practices. Each of these five steps contains a question that is intended to help you focus on a particular security system or security countermeasure. The questions may seem, at first, to be obvious, even trivial. But if you bear with me, and take them seriously, you will find they will help you determine which kinds of security make sense and which don't.

- *Step 1: What assets are you trying to protect?* This question might seem basic, but a surprising number of people never ask it. The question involves understanding the scope of the problem. For example, securing an airplane, an airport, commercial aviation, the transportation system, and a nation against terrorism are all different security problems, and require different solutions.
- *Step 2: What are the risks to these assets?* Here we consider the need for security. Answering it involves understanding what is being defended, what the consequences are if it is successfully attacked, who wants to attack it, how they might attack it, and why.
- *Step 3: How well does the security solution mitigate those risks?* Another seemingly obvious question, but one that is frequently ignored. If the security solution doesn't solve the problem, it's no good. This is not as simple as looking at the security solution and seeing how well it works. It involves looking at how the security solution interacts with everything around it, evaluating both its operation and its failures.
- *Step 4: What other risks does the security solution cause?* This question addresses what might be called the problem of unintended consequences. Security solutions have ripple effects, and most cause

new security problems. The trick is to understand the new problems and make sure they are smaller than the old ones.

- *Step 5: What costs and trade-offs does the security solution impose?*
  Every security system has costs and requires trade-offs. Most
  security costs money, sometimes substantial amounts; but other
  trade-offs may be more important, ranging from matters of convenience and comfort to issues involving basic freedoms like privacy. Understanding these trade-offs is essential.

These five steps don't lead to an answer, but rather provide the
mechanism to evaluate a proposed answer. They lead to another question: Is the security solution worth it? In other words, is the benefit of
mitigating the risks (Step 3) worth the additional risks (Step 4) plus
the other trade-offs (Step 5)? It is not enough for a security measure
to be effective. We don't have limitless resources or infinite patience.
As individuals and a society, we need to do the things that make the
most sense, that are the most effective use of our security dollar. But,
as you'll see later in this book, subjective (and sometimes arbitrary)
economic incentives make an enormous difference as to which security
solutions are cost-effective and which ones aren't.

These five steps may seem obvious when stated in this abstract
form, but applying them to real situations is hard work. Step 3, for
example, requires you to understand the security solution and how
well it actually works—not merely as described in the manufacturer's
literature or as explained to you by a politician, but in real situations
against real attackers. Step 5 requires you to understand the variety of
trade-offs a security solution imposes, how it interacts with other
aspects of your life, and what ancillary security solutions are required
to make it work. The answers aren't always easy to come by. It's hard
to quantify any of the steps, including the threat. Sometimes you won't
have enough information, and you will be forced to make your best
guess. Sometimes you'll be comparing apples and oranges. It can
quickly become very complex and subjective.

Keep the steps in mind as you read the many examples in this
book, and as you encounter and interact with the thousands of security
systems that surround you every day, you'll start to notice the difference between good and bad security or security that could be designed
better or require less onerous trade-offs. You'll begin to realize how
ineffectual some common security systems are: how they solve the
wrong problem or cause more problems than they solve. You'll start

making different security decisions: choosing more security in some instances and less security in others. You'll become more aware of the security trade-offs being imposed on you. And because security is all of a piece, the same knowledge will make you better able to choose a home alarm, and better equipped to participate in the national security policy debates.

Chapter 2

Security Trade-offs Are Subjective

There is no single correct level of security; how much security you have depends on what you're willing to give up in order to get it. This trade-off is, by its very nature, subjective—security decisions are based on personal judgments. Different people have different senses of what constitutes a threat, or what level of risk is acceptable. What's more, between different communities, or organizations, or even entire societies, there is no agreed-upon way in which to define threats or evaluate risks, and the modern technological and media-filled world makes these evaluations even harder.

There's no such thing as absolute security. It's human nature to wish there were, and it's human nature to give in to wishful thinking. But most of us know, at least intuitively, that perfect, impregnable, completely foolproof security is a pipe dream, the stuff of fairy tales, like living happily ever after. We have to look no further than the front page of the newspaper to learn about a string of sniper attacks, a shoe bomber on a transatlantic flight, or a dictatorial regime threatening to bring down a rain of fire on its enemies.

In Chapter 1, we talked about the fact that security always involves trade-offs. A government weighs the trade-offs before deciding whether to close a major bridge because of a terrorist threat; is the added security worth the inconvenience and anxiety? A homeowner weighs the trade-offs before deciding whether to buy a home alarm system; is the added security worth the bother and the cost? People make these sorts of security trade-offs constantly, and what they end up with isn't absolute security but something else. A few weeks after 9/11, a reporter asked me whether it would be possible to prevent a repetition of the terrorist attacks. "Sure," I replied, "simply ground all the aircraft." A totally effective solution, certainly, but completely impractical in a modern consumer-oriented society. And what's more, it would do nothing to address the general problem of terrorism, only certain specific airborne attacks.

Still, this extreme and narrowly effective option is worth considering. Ground all aircraft, or at least all commercial aircraft? That's exactly what did happen in the hours after the 9/11 attacks. The FAA ordered the skies cleared. Within two and a half hours, all of the 4,500 planes in the skies above this country were pulled down. Transcontinental flights were diverted to airports in Canada and elsewhere. Domestic flights landed at the closest possible airport. For the first time in history, commercial flight was banned in the U.S. In retrospect, this was a perfectly reasonable security response to an unprecedented attack. We had no idea if there were any other hijacked planes, or if there were plans to hijack any other planes. There was no quick and efficient way to ensure that every plane in the air was under the control of legitimate airline pilots. We didn't know what kind of airline security failures precipitated the attacks. We saw the worst happen, and it was a reasonable assumption that there was more to follow. Clearing the skies of commercial airplanes—even for only a few days—was an extreme trade-off, one that the nation would never have made in anything but an extreme situation.

Extreme trade-offs are easy. Want to protect yourself from credit card fraud? Don't own a credit card. Want to ensure that you are never mugged? Live on a deserted island. Want to secure yourself against mad cow disease? Never eat meat products. But if you want to eat prime rib in a crowded restaurant and pay with your credit card, you're going to have to accept imperfect security. Choosing to stay away from dark alleys may seem like a simple decision, but complex trade-offs lurk under the surface. And just as the U.S. cleared the skies in the wake of 9/11, someone who never worries about where he walks might change his mind if there's a crime wave in his neighborhood. All countermeasures, whether they're to defend against credit card fraud, mugging, or terrorism, come down to a simple trade-off: What are we getting, and what are we giving up in order to get it?

Most countermeasures that increase the security of air travel certainly cost us significantly more money, and will also cost us in terms of time and convenience, in almost any way that they might be implemented. A new security procedure for airplanes, one that adds a minute to the check-in time for each passenger, could add hours to the total preparation time for a fully booked 747. The government has not seriously considered barring all carry-on luggage on airplanes, simply because the flying public would not tolerate this restriction. Positive bag matching of passengers with their luggage was largely accomplished with information systems, but a thorough check of every passenger's carry-on bag would be another matter entirely; already many travelers have decided to drive instead of flying, where possible, because airline security has become so intrusive.

Much of our sense of the adverse, then, comes down to economics, but it also derives from some nebulous cultural sense of just how much disorder, or larceny, or invasion of privacy, or even mere rudeness we'll tolerate. For example, Swiss attitudes toward security are very different from American ones. The public toilet stalls in the Geneva airport are walled cubicles with full-length doors and better locks than on the doors of many U.S. homes. The cubicles take up a lot of space and must be a real nuisance to clean, but you can go inside with your luggage and nobody can possibly steal it.

The Swiss are willing to pay higher prices for better security in myriad ways. Swiss door locks are amazing; a standard apartment lock is hard to pick and requires a key that can't be duplicated with readily available equipment. A key can be duplicated only by the lock manufacturer at the written request of the property owner. The trade-offs of this are pretty significant. One of my colleagues had a minor lock problem, and his whole family was locked out for hours before it could be fixed. Many Swiss families have only one or two house keys, regardless of the number of people living together.

The security a retail store puts in place will depend on trade-offs. The store's insurance company might mandate certain security systems. Local laws might prohibit certain security systems or require others. Different security systems cost different amounts and disrupt normal store operations to different degrees. Some decisions about security systems are optional, and some are not.

Because security always involves a trade-off, more security isn't always better. You'd be more secure if you never left your home, but what kind of life is that? An airline could improve security by strip-searching all passengers, but an airline that did this would be out of business. Studies show that most shoplifting at department stores occurs in fitting rooms. A store can improve security by removing the fitting rooms, but many stores believe that the resulting decrease in profits from sales would be greater than the cost of shoplifting. It's the trade-off that's important, not the absolute security. The question is: How do you evaluate and make sensible trade-offs about the security in your own life?

. . . .

In many instances, people deliberately choose less security because the trade-offs for more security are too large. A grocery store might decide not to check expiration dates on coupons; it's not worth holding up the checkout line and inconveniencing good customers to catch the occasional fifty-cent fraud. A movie theater might not worry about people trying to sneak in while others are exiting; the minor losses aren't worth paying the extra guards necessary to forestall them. A retail shop might ignore the possibility that someone would lie about his age to get the senior citizen discount; preventing the fraud isn't worth asking everyone for proof of age.

In the professional lingo of those who work in security, there is an important distinction drawn between the words "threat" and "risk." A *threat* is a potential way an attacker can attack a system. Car burglary, car theft, and carjacking are all threats—in order from least serious to most serious (because an occupant is involved). When security professionals talk about *risk*, they take into consideration both the likelihood of the threat and the seriousness of a successful attack. In the U.S., car theft is a more serious risk than carjacking because it is much more likely to occur.

Most people don't give any thought to securing their lunch in the company refrigerator. Even though there's a threat of theft, it's not a significant risk because attacks are rare and the potential loss just isn't a big deal. A rampant lunch thief in the company changes the equation; the threat remains the same, but the risk of theft increases. In response, people might start keeping their lunches in their desk drawers and accept the increased risk of getting sick from the food spoiling. The million-dollar Matisse painting in the company's boardroom is a more valuable target than a bagged lunch, so while the threat of theft is the same, the risk is much greater. A nuclear power plant might be under constant threat both of someone pasting a protest sign to its front door and of someone planting a bomb near the reactor. Even though the latter threat is much less likely, the risk is far greater because the effects of such damage are much more serious.

Risk management is about playing the odds. It's figuring out which attacks are worth worrying about and which ones can be ignored. It's spending more resources on the serious attacks and less on the frivolous ones. It's taking a finite security budget and making the best use of it. We do this by looking at the risks, not the threats. Serious risks—either because the attacks are so likely or their effects are so devastating—get defended against while trivial risks get ignored. The goal isn't to eliminate the risks, but to reduce them to manageable levels. We know that we can't eliminate the risk of burglaries, but a good door lock combined with effective police can reduce the risk substantially. And while the threat of a paramilitary group attacking your home is a serious one, the risk is so remote that you don't even think about defending your house against it. The owner of a skyscraper will look at both attacks differently because the risk is different, but even she will ignore some risks. When in a million years would a 767 loaded with jet fuel crash into the seventy-seventh floor of a New York skyscraper? Threats determine the risks, and the risks determine the countermeasures.

Managing risks is fundamental to business. A business constantly manages all sorts of risks: the risks of buying and selling in a foreign currency, the risk of producing merchandise in a color or style the public doesn't want, the risk of assuming that a certain economic climate will continue. Threats to security are just another risk that a business has to manage. The methods a business employs in a particular situation depend on the details of that situation. Think, for example, of the different risks faced by a large grocery chain, a music shop, or a jewelry store.

There are many ways to deal with "shrinkage," as shoplifting is referred to in parts of the retail industry. Most chain grocery stores, for example, simply accept the risk as a cost of doing business; it's cheaper than installing countermeasures. A music shop might sell its CDs in bulky packaging, making them harder to shoplift. A jewelry store might go further and lock up all the merchandise, not allowing customers to handle any object unattended.

Then there's the matter of insurance. Of the three retail operations, probably only the jewelry store will carry theft insurance. Insurance is an interesting risk management tool. It allows a store to take its risk and, for a fee, pass it off to someone else. It allows the store to convert a variable-cost risk into a fixed-cost expense. Insurance reduces risk to the store owner because it makes part of the risk someone else's problem.

Like the U.S. government clearing the skies after 9/11, all of these solutions are situational. What a store does depends on what products it's selling, what its building is like, how large a store it is, what neighborhood it's in, what hours it's open, what kinds of customers it gets, whether its customers pay cash or buy using traceable checks or credit cards, whether it tends to attract one-time customers or repeat customers, et cetera.

These factors also affect the risk of holdups. A liquor store might decide to invest in bulletproof glass between the customer and the sales clerk and merchandise, with a specially designed window to pass bottles and money back and forth. (Many retail stores in south-central Los Angeles are like this.) An all-night convenience store might invest in closed-circuit cameras and a panic button to alert police of a holdup in progress. In Israel, many restaurants tack a customer surcharge onto their bills to cover the cost of armed guards. A bank might have bulletproof glass, cameras, and guards, and also install a vault.

Most of this is common sense. We don't use the same security countermeasures to protect diamonds that we use to protect donuts; the value and hence the risks are completely different.

And the security must balance the risk. What all of these businesses are looking for is to maximize profits, and that means adequate security at a reasonable cost. Governments are trying to minimize both loss of life and expenditures, and that also means adequate security at a reasonable cost. The decision to use guards is based not only on risk, but on the trade-off between the additional security and the cost of the guards. When I visited Guatemala City in the 1980s, all major stores and banks had armed guards. Guards cost less, per hour, in Guatemala than in the U.S., so they're more likely to be used. In Israel, on the other hand, guards are expensive, but the risks of attack are greater. Technological security measures are popular in countries where people are expensive to hire and the risks are lower.

Balancing risks and trade-offs are the point of our five-step process. In Step 2, we determine the risks. In Steps 3 and 4, we look for security solutions that mitigate the risks. In Step 5, we evaluate the trade-offs. Then we try to balance the pros and cons: Is the added security worth the trade-offs? This calculation is risk management, and it tells us what countermeasures are reasonable and what countermeasures are not.

Everyone manages risks differently. It's not just a matter of human imperfection, our inability to correctly assess risk. It also involves the different perspectives and opinions each of us brings to the world around us. Even if we both have the same knowledge and expertise, what might seem like adequate security to me might be inadequate to you because we have different tolerances for risk. People make value judgments in assessing risk, and there are legitimate differences in their judgments. Because of this fact, security is subjective and will be different for different people, as each one determines his own risk and evaluates the trade-offs for different countermeasures.

I once spoke with someone who is old enough to remember when a front-door lock was first installed in her house. She recalled what an imposition the lock was. She didn't like having to remember to take her key each time she went out, and she found it a burden, on returning home, to have to fumble around to find it and insert it in the lock, especially in the dark. All this fuss, just to get into her own home! Crime was becoming a problem in the town where she lived, so she understood why her parents had installed the lock, but she didn't want to give up the convenience. In a time when we've all become security-conscious, and some have become security-obsessed, an attitude like hers may seem quaint. Even so, I know people who still leave their doors unlocked; to them, the security trade-off isn't worth it.

It's important to think about not just the huge variety of security choices open to us, but the huge variety of personal responses to those choices. Security decisions are personal and highly subjective. Maybe a trade-off that's totally unacceptable to you isn't at all a problem for me. (You wouldn't consider living without theft insurance, whereas I couldn't be bothered.) Maybe countermeasures that I find onerous are perfectly okay with you. (I don't want my ID checked every time I enter a government building; you feel more secure because there's a guard questioning everyone.) Some people are willing to give up privacy and control to live in a gated community, where a guard (and, increasingly, a video system) takes note of each vehicle entering and exiting. Presumably, the people who live in such communities make a conscious decision to do so, in order to increase their personal security. On the other hand, there are those who most emphatically don't want their comings and goings monitored. For them, a gated community is anathema. There's not likely to be much in the way of a fruitful discussion of trade-offs between someone absolutely committed to planned community living and someone muttering about creeping encroachments on privacy. But the difference of opinion between the two (like the differences between those facing gun-control questions or workplace surveillance options) is just that—a valid difference of opinion.

When someone says that the risks of nuclear power (for example) are unacceptable, what he's really saying is that the effects of a nuclear disaster, no matter how remote, are so unacceptable to him as to make the trade-off intolerable. This is why arguments like "but the odds of a core meltdown are a zillion to one" or "we've engineered this reactor to have all sorts of fail-safe features" have no effect. It doesn't matter how unlikely the threat is; the risk is unacceptable because the consequences are unacceptable.

A similar debate surrounds genetic engineering of plants and animals. Proponents are quick to explain the various safety and security measures in place and how unlikely it is that bad things can happen. Opponents counter with disaster scenarios that involve genetically engineered species leaking out of the laboratories and wiping out other species, possibly even ours. They're saying that despite the security in place that makes the risk so remote, the risk is simply not worth taking. (Or an alternative argument: that the industry and the government don't understand the risks because they overestimate the efficacy of their security measures, or in any event can't be trusted to properly take them into account.)

For some people in some situations, the level of security is beside the point. The only reasonable defense is not to have the offending object in the first place. The ultimate example of this was the speculation that the Brookhaven National Lab's proposed ion collider could literally destroy the universe.

Sometimes perceptions of unacceptable risk are based on morality. People are unwilling to risk certain things, regardless of the possible benefits. We may be unwilling to risk the lives of our children, regardless of any rational analysis to the contrary. The societal version of this is "rights"; most people accept that things like torture, slavery, and genocide are unacceptable in civilized society.

For some, the risks of some attacks are unacceptable, as well: for example, a repetition of 9/11. Some people are willing to bear any cost to ensure that a similar terrorist attack never occurs again. For others, the security risks of visiting certain countries, flying on airplanes, or enraging certain individuals are unacceptable. Taken to the extreme, these fears turn into phobias. It's important to understand that these are personal, and largely emotional, reactions to risk. The risks can be wildly unlikely, but they are important nonetheless because people act on their perceptions.

Even seemingly absolute risk calculations may turn out to be situational. How far can you push the activist who is fearful of a runaway genetic modification? What would the activist say, for example, if the stakes were different—if a billion people would starve to death without genetically modified foods? Then the risks might be acceptable after all. (This calculation has repeatedly occurred in Africa in recent years, with different countries making different decisions. The best compromise seems to be accepting genetically modified foodstuffs that can't replicate: imported flour but not imported grain. But some famine-stricken countries still reject genetically modified flour.) In the months after 9/11, perfectly reasonable people opined that torture was acceptable in some circumstances. Think about the trade-offs made by the people who established the Manhattan Project: Having fission bombs available to the world might be risky, but it was felt that the bombs were less risky than having the Nazis in control of Europe.

These calculations are not easy. There is always an imprecision, and sometimes the uncertainty factor is very large. Even with sufficient information, the calculation involves determining what is worth defending against and what isn't, or making a judgment about what losses are acceptable. It sometimes even involves understanding that a decision will result in some deaths, but that the alternatives are unreasonable. What is the risk of another 9/11-like terrorist attack? What is the risk that Al Qaeda will launch a different, but equally deadly, terrorist attack? What is the risk that other terrorist organizations will launch a series of copycat attacks? As difficult as these questions are, it is impossible to intelligently discuss the efficacy of antiterrorism security without at least some estimates of the answers. So people make estimates, or guess, or use their intuition.

. . . .

In the months after 9/11, a story circulated on the Internet about an armed air marshal who was allowed on board a plane with his firearm, but had to surrender his nail clippers. In January 2002, a US Airways pilot at the Philadelphia airport was hauled off to jail in handcuffs after he asked a screener, "Why are you worried about tweezers when I could crash the plane?" The pilot's remark was certainly ill-advised, but the reason so many laugh at the thought of tweezers being confiscated at airports is because we know that there is no increased risk in allowing them on board.

Most of us have a natural intuition about risk. We know that it's riskier to cut through a deserted alley at night than it is to walk down a well-lighted thoroughfare. We know that it's riskier to eat from a grill cart parked on the street than it is to dine in a fine restaurant. (Or is it? A restaurant has a reputation to maintain, and is likely to be more careful than a grill cart that disappears at the end of the day. On the other hand, at a cart I can watch my meat being cooked.) We have our own internal beliefs about the risks of trusting strangers to differing degrees, participating in extreme sports, engaging in unprotected sex, and undergoing elective surgery. Our beliefs are based on lessons we've learned since childhood, and by the time we're adults, we've experienced enough of the real world to know—more or less—what to expect. High places can be dangerous. Tigers attack. Knives are sharp.

A built-in intuition about risk—engendered by the need to survive long enough to reproduce—is a fundamental aspect of being alive. Every living creature, from bacteria on up, has to deal with risk. Human societies have always had security needs; they are as natural as our needs for food, clothing, and shelter. My stepdaughter's first word was "hot." An intuition about risk is a survival skill that has served our species well over the millennia.

But saying that we all have these intuitions doesn't mean, by any stretch, that they are accurate. In fact, our perceived risks rarely match the actual risks. People often underestimate the risks of some things and overestimate the risks of others. Perceived risks can be wildly divergent from actual risks compiled statistically. Consider these examples:

- People exaggerate spectacular but rare risks and downplay common risks. They worry more about earthquakes than they do about slipping on the bathroom floor, even though the latter kills far more people than the former. Similarly, terrorism causes far more anxiety than common street crime, even though the latter claims many more lives. Many people believe that their children are at risk of being given poisoned candy by strangers at Halloween, even though there has been no documented case of this ever happening.
- People have trouble estimating risks for anything not exactly like their normal situation. Americans worry more about the risk of mugging in a foreign city, no matter how much safer it might be than where they live back home. Europeans routinely perceive the U.S. as being full of guns. Men regularly underestimate how risky a situation might be for an unaccompanied woman. The risks of computer crime are generally believed to be greater than they are, because computers are relatively new and the risks are unfamiliar. Middle-class Americans can be particularly naïve and complacent; their lives are incredibly secure most of the time, so their instincts about the risks of many situations have been dulled.

- Personified risks are perceived to be greater than anonymous risks. Joseph Stalin said, "A single death is a tragedy, a million deaths is a statistic." He was right; large numbers have a way of blending into each other. The final death toll from 9/11 was less than half of the initial estimates, but that didn't make people feel less at risk. People gloss over statistics of automobile deaths, but when the press writes page after page about nine people trapped in a mine—complete with human-interest stories about their lives and families—suddenly everyone starts paying attention to the dangers with which miners have contended for centuries. Osama bin Laden represents the face of Al Qaeda, and has served as the personification of the terrorist threat. Even if he were dead, it would serve the interests of some politicians to keep him "alive" for his effect on public opinion.

- People underestimate risks they willingly take and overestimate risks in situations they can't control. When people voluntarily take a risk, they tend to underestimate it. When they have no choice but to take the risk, they tend to overestimate it. Terrorists are scary because they attack arbitrarily, and from nowhere. Commercial airplanes are perceived as riskier than automobiles, because the controls are in someone else's hands—even though they're much safer per passenger mile. Similarly, people overestimate even more those risks that they can't control but think they, or someone, should. People worry about airplane crashes not because we can't stop them, but because we think as a society we should be capable of stopping them (even if that is not really the case). While we can't really prevent criminals like the two snipers who terrorized the Washington, DC, area in the fall of 2002 from killing, most people think we should be able to.

- Last, people overestimate risks that are being talked about and remain an object of public scrutiny. News, by definition, is about anomalies. Endless numbers of automobile crashes hardly make news like one airplane crash does. The West Nile virus outbreak in 2002 killed very few people, but it worried many more because it was in the news day after day. AIDS kills about 3 million people per year worldwide—about three times as many people each day as died in the terrorist attacks of 9/11. If a lunatic goes back to the office after being fired and kills his boss and two co-workers, it's national news for days. If the same lunatic shoots his ex-wife and two kids instead, it's local news ... maybe not even the lead story.

In America, automobiles cause 40,000 deaths every year; that's the equivalent of a full 727 crashing every day and a half—225 total in a year. As a society, we effectively say that the risk of dying in a car crash is worth the benefits of driving around town. But if those same 40,000 people died each year in fiery 727 crashes instead of automobile accidents, you can be sure there would be significant changes in the air passenger systems. (I don't mean to harp on automobile deaths, but riding in a car is the riskiest discretionary activity the majority of Americans regularly undertake.) Similarly, studies have shown that both drivers and passengers in SUVs are more likely to die in accidents than those in compact cars, yet one of the major selling points of SUVs is that the owner feels safer in one.

This example illustrates the problem: People make security decisions based on perceived risks instead of actual risks, and that can result in bad decisions.

. . . .

A lot of this disconnect between actual risks and perceived risks is psychological, and psychologists and sociologists have paid a great deal of attention to it. And the problem is not getting better. Modern society, for all its undeniable advances, has complicated and clouded our ability to assess risk accurately in at least two important ways.

The first is the rapid advance of technology. Twenty generations ago, people lived in a society that rarely changed. With a few notable exceptions, the social and economic systems they were born into were the systems in which they and their parents and grandparents spent their entire lives. In these societies, people could go through virtually their entire adult lives without hearing a sound they'd never heard before or seeing a sight they'd never seen before. Only a privileged few (and soldiers and sailors) traveled very far from the place of their birth. And changes—someone would invent gunpowder, the stirrup, or a better way of building an arch—were slow to come, and slow to be superseded. People learned how to live their lives, and what they learned was likely to serve them well for an entire lifetime. What was known and trusted was known and trusted for generations.

Nowadays every day seems to bring a new invention, something you've not seen before. Of course the large things about the world—concepts, interactions, institutions—are relatively constant, but a lot of the details about it are in constant flux. Cars, houses, and bank accounts all have features they didn't have ten years ago, and since most people don't understand how all these new features work, they can't evaluate the risks properly.

This phenomenon is ubiquitous. The average computer user has no idea about the relative risks of giving a credit card number to a Web site, sending an unencrypted e-mail, leaving file sharing enabled, or doing any of the dozens of things he does every day on the Internet. The average consumer doesn't understand the risks involved in using a frequent shopper ID card at his local supermarket, paying with a check, or buying genetically engineered food. People can easily read about—in fact, they can hardly avoid reading about—the risks associated with stock manipulation, biological terror weapons, or laws giving police new powers. But this does not mean that they understand, or are capable of managing, these risks. Technological progress is now faster than our ability to absorb its implications. I am reminded of stories of farmers from the countryside coming to the big city for the first time. We are all rubes from the past, trying to cope with the present day.

And it is this change coming to the farmer, more than the farmer coming to the city, that is the second challenge presented by modern society. Modern mass media, specifically movies and TV news, has degraded our sense of natural risk. We learn about risks, or we think we are learning, not by directly experiencing the world around us and by seeing what happens to others, but increasingly by getting our view of things through the distorted lens of the media. Our experience is distilled for us, and it's a skewed sample that plays havoc with our perceptions. Kids try stunts they've seen performed by professional stuntmen on TV, never recognizing the precautions the pros take. The five o'clock news doesn't truly reflect the world we live in—only a very few small and special parts of it.

Slices of life with immediate visual impact get magnified; those with no visual component, or that can't be immediately and viscerally comprehended, get downplayed. Rarities and anomalies, like terrorism, are endlessly discussed and debated, while common risks like heart disease, lung cancer, diabetes, and suicide are minimized.

The global reach of today's news further exacerbates this problem. If a child is kidnapped in Salt Lake City during the summer, mothers all over the country suddenly worry about the risk to their children. If there are a few shark attacks in Florida—and a graphic movie—suddenly every swimmer is worried. (More people are killed every year by pigs than by sharks, which shows you how good we are at evaluating risk.)

Deaths in the United States, per Year, Due to Various Causes

Heart disease	725,192
Cancer	549,838
Diabetes	68,399
Flu and pneumonia	63,730
Motor vehicle accidents	41,700
Murder	15,586
HIV	14,478
Food poisoning	5000
Residential fires	3465
Upholstered furniture (catching on fire)	693
Plane accidents	631
Train accidents	530
Floods	139
Deer (collisions with vehicles)	135
Lightning	87
Tornadoes	82
Nursery product accidents (cribs, cradles, etc.)	63
Hurricanes	27
Playground equipment accidents	22
Dogs	18
Snakes	15
Mountain lions	0.7
Sharks	0.6

Note: This data has been compiled from a variety of sources, and some numbers may be more accurate than others. The point here is not to exactly specify the actual risks, but to illustrate that life is filled with unexpected risks, and that the risks people worry about are rarely the most serious ones.

Movies also skew our intuition by presenting unreal scenarios. We are led to believe that problems can be solved in less than two hours, and that the hero can both stop the bad guys and get the girl in the final ten minutes. Movies show technology delivering miracles and lone heroes defeating intricate security systems: James Bond, *Mission Impossible*, and so on. People believe that the world is far more intricate and devious than it really is. The effects can be seen in courtroom juries, who are more willing to believe a labyrinthine conspiracy theory than a much more likely straightforward explanation. "Conspiracy theories are an irresistible labor-saving device in the face of complexity," says Harvard professor H.L. "Skip" Gates.

All this has been true since the beginning of civilization—much narrative is built from heroic stories—but it has never been as pervasive and realistic as today. As people's ability to accurately determine risks falters even more, when their natural intuition isn't sufficient, they fall back on fear, misconceptions, and even superstitions. The ramifications have profound implications on security. Because we do not understand the risks, we make bad security trade-offs.

Chapter 3

Security Trade-offs Depend on Power and Agenda

Most security decisions are complicated, involving multiple players with their own subjective assessments of security. Moreover, each of these players also has his own agenda, often having nothing to do with security, and some amount of power in relation to the other players. In analyzing any security situation, we need to assess these agendas and power relationships. The question isn't which system provides the optimal security trade-offs—rather, it's which system provides the optimal security trade-offs for which players.

One of the reasons security is so hard to get right is that it inevitably involves different parties—let's call them *players*—each with his or her subjective perceptions of risk, tolerances for living with risk, and willingnesses to make various trade-offs. It should come as no surprise, then, that there is a strong tendency for a player involved in a security system to approach security subjectively, making those trade-offs based on both his own analysis of the security problem and his own internal and external non-security considerations: collectively, his *agenda*.

Think back to the week after 9/11, and imagine that all the players involved in airline security were in a room trying to figure out what to do. Some members of the public are scared to fly (each person to his own degree) and need to be reassured that everything is going to be okay. The airlines are desperate to get more of the public flying but are leery of security systems that are expensive or play havoc with their flight schedules. They are happy to let the government take over the job of airport screening because then it won't be their fault if there's a problem in the future. Many pilots like the idea of carrying guns, as they now fear for their lives. Flight attendants are less happy with the idea, afraid that they could be left in danger while the pilots defend themselves. Elected government officials are concerned about reelection and need to be seen by the public as doing *something* to improve security. And the FAA is torn between its friends in the airlines and its friends in government. Confiscating nail files and tweezers from passengers seems like a good idea all around: The airlines don't mind because it doesn't cost them anything, and the government doesn't mind because it looks like it's doing something. The passengers haven't been invited to comment, although most seasoned travelers simply roll their eyes.

As a security expert reviewing this imaginary scenario, I am struck by the fact that no one is trying to figure out what the optimal level of risk is, how much cost and inconvenience is acceptable, and then what security countermeasures achieve these trade-offs most efficiently. Instead, everyone is looking at the security problem from his or her own perspective.

And there are many more players, with their own agendas, involved in airline security. Did you ever wonder why tweezers were confiscated at security checkpoints, but matches and cigarette lighters—actual combustible materials—were not? It's because the tobacco lobby interjected its agenda into the negotiations by pressuring the government. If the tweezers lobby had more power, I'm sure they would have been allowed on board, as well. Because there are power imbalances among the different parties, the eventual security system will work better for some than for others.

A security system implies a *policy* of some sort, which in turn requires someone who defines or has defined it. Going back to our basic definition, the policy will be aimed at "preventing adverse consequences from the intentional and unwarranted actions of people." This definition presupposes some kind of prior relationship between the players involved, specifically an agreement on what "unwarranted actions" are. In every instance of security, someone—generally the asset owner—gets to define what is an unwarranted action and what is not, and everyone else is obliged to go along with that definition. All security can be—in fact, needs to be—studied in terms of agendas defining policy, with the absolutely inevitable consequence that different players gain and lose as a result.

It takes two players to create a security problem: the attacker and the attacked (the asset owner). If there's no attacker, then there's obviously no security problem (although the perception of an attacker may be enough). If no one is attacked, there's no security problem either. Even theft is not necessarily a security problem if it does no harm and the owner is not discomforted; in other words, the attack has to result in what's agreed on as an "adverse consequence." Someone who walks around a city park picking up litter is not a security problem. The same person walking around picking flowers, however, might be a security problem. And once she starts picking pockets, she's definitely a security problem. Policies, then, may be codified into law, but can also involve the unspoken, habitual, traditional, unwritten, and social. Personal security policies are more driven by societal norms than by law. Business security policies are a result of a company's self-interest.

Self-interest has profound effects on the way a player views a security problem. Except for the inconvenience, credit card fraud is not much of a security problem to the cardholder, because in the U.S., the banks shoulder all but $50 of the liability. That $50 is a token liability: enough to make you concerned about the loss, but not so much that you'd be afraid to carry your credit card with you when you shop. Change the individual cardholder's liability to $500, for example, and his attitude toward the seriousness of the credit card fraud problem will certainly change overnight. It might even call into question the value of having a credit card.

Security systems are never value-neutral; they move power in varying degrees to one set of players from another. For example, anti-counterfeiting systems enforce the government's monopoly on printing money, taking that ability away from criminals. Pro-privacy technologies give individuals power over their personal information, taking that power away from corporations and governments. In some systems, such as anti-counterfeiting countermeasures, the security system simply works to support a codified power arrangement—that is, one enshrined in law. Other security systems are open, dynamic, unresolved: Privacy countermeasures like paying for purchases with anonymous cash instead of a credit card have no legal precedents, and instead are part of an ongoing power struggle between people on one side, and corporations and governments on the other. But in both cases, the security systems are part of a greater social system.

Sometimes a policy is straightforward, particularly if it applies to a relatively small and easy-to-define unit. I control the basic security policy of my house. I decide who is allowed in, who gets a key, and when the doors and windows are locked. (The security countermeasures I choose may be more or less effective in enforcing that security policy, but that's a different story.) A corporation gets to control its security—who is permitted access to different buildings and offices and supply closets, who keeps the company's books, who signs contracts in the company's name—and delegates that control to individual employees. That's a somewhat more elaborate security policy, but still it is fairly easy to plan, implement, and manage.

Other security systems, of course, enforce a much more complex policy. The credit card system involves many players: the customer, the merchant, the bank that issued the credit card, the merchant's bank as it tries to collect the money, the credit card companies, the government. All of these players have different security needs and concerns about the system, and the security countermeasures will protect them all to differing degrees. In this system, the players range from individuals and groups of individuals to institutional players. Some of these institutions are themselves highly complex hierarchies and have significant shared and opposing needs. But even a world-wide credit card operation will have relatively simple needs when compared to national security. Here the players again range from the individual to the institutional, with complex and varied needs, wants, concerns, hopes, and fears. Securing your home is a much simpler task than securing your nation, partly because in the latter case there are so many different players.

....

The notion of agendas is further complicated when one or more of the parties is represented by a proxy. *Proxies* are players who act in the interest of other players. As society has gotten more complex and technological, individuals have created proxies to take over the task of risk management and to provide them with some level of security. Unfortunately, the proxy is not simply the amalgamation of the will of the people; it is, in and of itself, a new player with an agenda that may not match—in fact will rarely exactly match—the agenda of the person it represents.

Most people have no idea how to evaluate the safety of airplanes or prescription drugs. They don't have the time, the information, or the expertise. They can't possibly manage their own risks. In these cases and others, the government steps in as a proxy. Through different regulatory agencies—in the U.S., the FAA and FDA are examples—the government acts as an advocate for the consumer, subject, of course, to political and commercial pressures. Most people have no idea about the intricacies of the legal system. Instead of having to navigate them alone, they hire lawyers to act as their proxies. Most people don't know anything about how the plumbing in their home works. Instead of becoming experts in the subject, they hire commercial plumbers to take care of it for them. In a modern republic, elected officials are citizens' proxies in the government's decision-making process.

Proxies are a natural outgrowth of society, an inevitable by-product of specialization—everyone can't do everything, let alone master the ongoing technical complexity of modern life. This phenomenon has profound effects on security, greater than any individual security technology or countermeasure or system. Proxies are not necessarily going to make the same risk management decisions that the people they represent would make. They're not necessarily going to make the same trade-offs between functionality and security, between personal freedom and security, between cost and security, as the people would. They're going to bring their own agendas into the security debate. This fact determines not only how well security systems are implemented, but also which security systems are even considered. It determines which power relationship the security serves to enforce, and which players the security measures benefit. And it determines how effectively a security countermeasure will be employed and how often it will be bypassed.

For example, before I bought my home, I hired a building inspector. I don't have the expertise to evaluate the physical condition of a house, so I hired a proxy to protect me from making a bad purchase. My real estate agent suggested the building inspector. I don't plan on buying another house for decades; the real estate agent sells dozens of houses a year. Conflict of interest inheres in this system. Even though the inspector was theoretically acting in my interests, he needs referrals from the real estate agent more than he needs me. Even though he is my proxy, his interests aren't the same as mine. His actual job is to convince me to buy the house. He's not going to find all kinds of reasons why I shouldn't, because if he did that consistently, the agent would stop recommending him. In my case this turned out not to be a problem, but I have friends who believe that their building inspectors minimized the seriousness of some things and ignored other things—just so the sale would go through.

Conflicts of interest are not, of course, unique to security. Companies do not have the same agenda as their employees, their customers, or even their chief executive or their board of directors. Inevitably, organizations develop institutional agendas of their own, as do departments within those organizations. Conflicts of interest can also be seen in elected officials, who do not have agendas identical to the people who vote for them, and police officers, who do not have agendas identical to those of the people they're entrusted to protect. These groups act in their own self-interest—companies protect their own bottom line, elected officials enrich their power and their ability to get reelected, police officers protect their institution—while at the same time trying to do their jobs as proxies. While there are different degrees of congruence, proxies—government agencies, courts, insurance companies, and independent testing laboratories—are not identical to the people who turn to them.

. . . .

A player's agenda is about more than security, and often non-security concerns trump security. In Chapter 1, I wrote that security is partially a state of mind. If this is true, then one of the goals of a security countermeasure is to provide people with a feeling of security in addition to the reality. But some countermeasures provide the feeling of security *instead of* the reality. These are nothing more than *security theater*. They're palliative at best.

In 1970, there was no airline security in the U.S.: no metal detectors, no X-ray machines, and no ID checks. After a hijacking in 1972—three men took over a plane and threatened to crash it into the Oak Ridge, Tennessee, nuclear power plant—airlines were required to post armed guards in passenger boarding areas. This countermeasure was less to decrease the risk of hijacking than to decrease the anxiety of passengers. After 9/11, the U.S. government posted armed National Guard troops at airport checkpoints primarily for the same reason (but were smart enough not to give them bullets). Of course airlines would prefer it if all their flights were perfectly safe, but actual hijackings and bombings are rare events whereas corporate earnings statements come out every quarter. For an airline, for the economy, and for the country, judicious use of security theater calmed fears . . . and that was a good thing.

Tamper-resistant packaging is also largely a piece of security theater. It's easy to poison many foods and over-the-counter medicines right through the seal—by using a syringe, for example—or to open and replace the seal well enough that an unwary consumer won't detect it. But product poisonings are very rare, and seals make the buying public feel more secure. Sometimes it seems those in charge—of governments, of companies—need to do *something* in reaction to a security problem. Most people are comforted by action, whether good or bad.

There are times when a player creates compelling security theater precisely because its agenda doesn't justify real security. The cell phone industry isn't really interested in providing security against eavesdropping; it's not worth the trade-offs to them. What they are really interested in providing is security against fraud, because that directly affects the companies' bottom lines. Voice privacy is just another attractive feature, as long as it doesn't affect performance or phone size. As a result, we see a whole lot of advertising about the "natural security" of digital cellular—mostly nonsense, by the way—which results in a misplaced belief in security but no real privacy features. Even though the trade-off might be worth it to customers, most customers aren't good judges of cell phone security because the technology is simply too complicated. Instead, they have to rely on the phone companies. Offering security theater can improve market share just as much as offering actual security, and it is significantly cheaper to provide.

Comparing this to battery security is another story entirely. Nokia spends about a hundred times more money per phone on battery security than on communications security. The security system senses when a consumer uses a third-party battery and switches the phone into maximum power-consumption mode; the point is to ensure that consumers buy only Nokia batteries. Nokia is prepared to spend a considerable amount of money solving a security problem that it perceives—it loses revenue if customers buy batteries from someone else—even though that solution is detrimental to consumers. Nokia is much less willing to make trade-offs for a security problem that consumers have.

Other times, a player creates more security than is warranted because of a need for security theater. When Tylenol's manufacturer modified the product's packaging in the wake of the 1982 random poisonings, it did so because it needed to restore consumer confidence in its products. Were the additional packaging costs worth the minimal increase in security? I doubt it, but the increase in packaging costs was definitely worth the restored sales due to a reassured consumer base. The market dictated what the manufacturer should spend on increasing the feeling of security, even before any government regulations got into the act. "Tamperproof packaging" became good marketing.

It's important to understand security theater for what it is, but not to minimize its value. Security theater scares off stupid attackers and those who just don't want to take the risk. And while organizations can use it as a cheaper alternative to real security, it can also provide substantial non-security benefits to players. If you were living in Washington, DC, while the 2002 snipers were loose and your daughter was scared to walk home from school, you might have decided to drive her instead. A security expert could calculate that the increased security from your actions were negligible and were more than negated by the increased likelihood of getting into a car accident. But if driving her home from school made her better able to sleep at night, then buying into a piece of security theater was worth it.

. . . .

This notion of agenda—personal and corporate and bureaucratic—explains a lot about how security *really* works in the real world, as opposed to how people might expect, and wish, it to work. For example:

- Think about the money in your wallet. You have an overriding agenda to be able to spend your money and therefore have a powerful vested interest in believing that the money you have is not counterfeit. A security system that relies on you checking for counterfeit money won't work because you won't do your part. Preventing counterfeiting is the government's agenda, and the government is the one that needs to expend effort to detect and combat forgery.
- When ATM cardholders in the U.S. complained about phantom withdrawals from their accounts, the courts generally held that the banks had to prove fraud. Hence, the banks' agenda was to improve security and keep fraud low, because they paid the costs of any fraud. In the UK, the reverse was true: The courts generally sided with the banks and assumed that any attempts to repudiate withdrawals were cardholder fraud, and the cardholder had to prove otherwise. This caused the banks to have the opposite agenda; they didn't care about improving security, because they were content to blame the problems on the customers and send them to jail for complaining. The result was that in the U.S., the banks improved ATM security to forestall additional losses—most of the fraud actually was not the cardholder's fault—while in the UK, the banks did nothing.
- The airline industry has a long history of fighting improvements in airplane safety. Treaties limit the amount of damages airlines had to pay families of international airplane crash victims, which artificially changed the economics of airplane safety. It actually made more economic sense for airlines to resist many airplane safety measures, and airplane safety improvements came only after airplane manufacturers received military development contracts and because of new government regulation. Notice that the

agenda of governments—increased passenger safety—was forced onto the airlines, because the governments had the power to regulate the industry.

- Bureaucracies have their own internal agendas, too. Most of the time, they're unnaturally conservative. They might not deliberately prevaricate, but a you-can't-get-fired-for-saying-no attitude might prevail. Recent news reports have talked about FBI field offices having different agendas from the FBI in Washington. On the other hand, many groups have a cover-your-ass mentality. When the DC area snipers were still at large in 2002, many school districts canceled outdoor events even though the risk of attack was minimal. Some went so far as to cancel school. They didn't want to be seen as the ones who ignored the risk if the worst occurred.

- Through 2002, the U.S. government tried to convince corporations to improve their own security: at nuclear power plants, chemical factories, oil refineries, software companies, and so on. Officials appealed to the CEOs' sense of patriotism, reminding them that improving security would help their country. That this had little real effect should surprise no one. If the CEO of a major company announced that he was going to reduce corporate earnings by 25 percent to improve security for the good of the nation, he would almost certainly be fired. If I were on the company's board of directors, I would fire him. Sure, the corporation has to be concerned about national security, but only to the point where its cost is not substantial.

- Sometimes individual agendas are a harrowing matter of life and death: On 1 September 1983, Korean Airlines Flight 007, on its way from Anchorage, Alaska, to Seoul, Korea, carrying 269 passengers and crew, strayed off its intended course and entered into Soviet airspace. It was destroyed in midair because the Soviet general in charge of air defense knew that the last time a plane violated his nation's airspace, the general in charge that night was shot. The general didn't care about getting the risks right, whether the plane was civilian or military, or anything else. His agenda was his own neck.

In all of these stories, each player is making security trade-offs based on his own subjective agenda, and often the non-security concerns are the most important ones. What this means is that you have to evaluate security opinions based on the positions of the players.

When a pharmaceutical manufacturer says that tamper-resistant packaging doesn't add much to security and would be inconvenient for the consumer, it's because the manufacturer does not want to pay for the security countermeasure. Tamper-resistant packaging is not worth the expense *to that company*. When software industry lobbying groups say that applying liability laws to software would not improve software quality and would be a disaster for the public, it's because the affected companies don't want the expense. Liability laws are not worth the expense to *them*. When the U.S. government says that security against terrorism is worth curtailing individual civil liberties, it's because the cost of that decision is not borne by those making it. Extra security is worth the civil liberty losses because *someone else* is going to suffer for it. Security decisions are always about more than security. Understanding a player's negotiating position requires you to understand his personal agenda and trade-offs.

In economics, what is called an externality occurs when one player makes a decision that affects another player, one not involved in the decision. It's the kind of problem that surfaces when a company can save substantial money by dumping toxic waste into the river, and everyone in the community suffers because of contaminated water. The community is the player that gets stuck with the externality because it is not involved in the company's decision. In terms of the overall good to society, it is a bad decision to dump toxic waste into the river. But it's a good decision for the company because it doesn't bear the cost of the effects. Unless you understand the players and their agendas, you will never understand why some security systems take the shape they do.

This is the way security works. At the foundation there is a security system protecting assets from attackers. But that security system is based on a policy defined by one or more of the players (usually the asset owner) and the perceived risks against those assets. The policy is also affected by other players and other considerations, often having nothing to do with security. The whole process is situational, subjective, and social. Understanding, and working with, these various agendas is often more important than any technical security considerations.

It's pointless to hope that a wave of selfless, nonsubjective security sensibilities will suddenly sweep across society. The agendas of, and the power relationships between, players are an inevitable part of the process; to think otherwise is to delude yourself. Security will always be a balancing game between various players and their agendas, so if you want a level of security that matches *your* agenda and *your* ideas of risk management, you're going to have to approach it as a social problem. Don't underestimate, or gloss over, the differences among players. Understand their different agendas, and learn how to work with them. Understand that the player with the power will get more of his agenda codified into the eventual security system. And remember that security issues will often get trumped by non-security ones.

Part Two

How Security Works

Chapter 4

Systems and How They Fail

Security is never simple. It's more than a series of countermeasures—security is a complex system that interacts with itself, the assets being protected, and the surrounding environment. These interactions affect security in a profound way. They are the points at which the system fails when attacked, and they are the points where the system fails even in the absence of attackers. Because security systems are designed to prevent attack, *how* the systems fail is critical. And because attackers are generally rarer than legitimate users, how the systems fail in the absence of attackers is generally more important than how they fail in the presence of attackers.

In thinking about security, most people will, sensibly enough, begin by considering specific attacks and defenses. "I need to secure the bank's money against robbers, so I will install a vault." But a vault is no more than a 6-inch-thick steel-and-concrete-composite wall between piles of cash and dangerous bank robbers. For the vault to be an effective security countermeasure, a lot of other factors must be taken into account. Just for starters, who knows the combination? What happens when she's fired, or quits, or just fails to show up for work? What happens if she dies? Who moves money in and out of the vault? When and how? How is the vault secured when the door is open? Who checks to see if the amount of money in the vault is the same as what bank personnel *think* should be in the vault? How often is this checked? What happens if there is a discrepancy? Are there safe-deposit boxes in the same vault? How do customers get to those boxes? Does it matter to the bank what customers put in their boxes? And who installed the vault? Does the installer know the combination, too? Do some of his employees?

Are there alarms on the vault? Who installed the alarm, and who responds when it rings? Who gets to decide if it's a real bank robbery or a false alarm? What happens if it's a real bank robbery? What happens if it's a false alarm? What happens if the responder can't tell the difference?

Questions, it seems, lead to ever more questions. What happens if the lock on the vault breaks and can't be opened? What happens if the lock breaks and the door can't be closed? What happens if there's an earthquake and the vault's walls crack? What happens if, as it did to the Bank of Nova Scotia on 11 September 2001, the bank is buried under 110 stories of collapsed building?

Isn't anything simple anymore?

No, at least not when it comes to security. The questions proliferate, and inevitably issues surrounding what at first seems little more than a big heavy box with a lock branch out to questions of personnel, and rules of access, and a myriad of other considerations. As some engineers say, the issues ramify. Faced with even a fairly routine security problem—how to protect the money in a bank—a security planner immediately has to deal not just with the what, but with the who and with the how—and all their interactions. Often security is less concerned about the assets being protected, and more about the functionality of those assets. What matters is the opening and closing of the safe, the taking money in and out, the installation, et cetera. To put it another way, adding security to anything requires a *system*, and systems are complex and elusive and potentially maddening beasts. But if you want to understand security, there is no escaping them. You are forced to think in terms of systems.

At the most basic level, a system is a collection of simpler components that interact to form a greater whole. A machine is a simple thing, even though it may have different pieces. A hammer is a machine; a table saw is a system. A pulley is a machine; an elevator is a system. A tomahawk is a machine; a Tomahawk cruise missile is a complex system.

And it gets even more complicated. Systems interact with other systems, forming ever-larger systems. A cruise missile is made up of many smaller systems. It interacts with launching systems and ground-control systems, which in turn interact with targeting systems and other military systems and even political systems. A building's elevator interacts with the building's electrical system, its fire-suppression system, its environmental control system, and its communications system, and, most important, it interacts with people. An aircraft carrier, the world's most complicated single manufactured product, is itself a system, containing all of these systems and many more besides.

Systems mark the difference between a shack and a modern building, a guard dog and a centralized alarm system, a landing strip and an airport. Anyone can build a stop sign—or even a traffic light—but it takes a different mind-set entirely to conceive of a citywide traffic control system. U.S. Navy Admiral Grace Hopper, a pioneering advocate of computer systems and information technologies, perhaps said it best: "Life was simple before World War II. After that, we had systems." Without systems—and our ability to combine smaller systems into larger systems, as well as to break larger systems down into smaller systems to better understand and engineer them—the complexity of modern-day life would be impossible.

The word *system* is also used to describe complex social, political, and economic processes—more specifically, collections of interacting processes. Going to a doctor is a process; health care is a system. Sitting down to dinner with your children is a process; a family is a system. Deciding how you live your own life is a process; deciding how other citizens live their lives is a political system. The commercial air transport system is a collection of smaller physical systems—airplanes, airports, a labor force—and smaller abstract systems like air traffic control and ticketing.

Even though the notion of a system is a modern invention, we can look back through history and find systems in all sorts of social arrangements: the Roman political system, the system of a medieval castle, and the Aztec agrarian system. Planting a field is an activity; feeding the population of Tenochtitlán required a system. Four walls and a roof make a hut. Put several huts together and populate them, and you've got a village—another system.

· · · ·

Security is a system, too—a system of individual countermeasures and the interactions between them. As such, it shares the traits of other systems. Take the bank vault example that opened this chapter. A modern vault, with its complex locking mechanism, is itself a system, but there's more. When a bank installs a vault, it also institutes systems of operation, auditing, and disaster recovery, to name a few. The vault is inextricably bound to these other systems. The security system is in itself a collection of assets plus functionality, and it affects the functionality of the very assets that are being protected.

Interactions between systems are inevitable, as we've discussed. But so far most of the interactions we've described are intentional; they're planned and designed in. It's axiomatic among system designers, though, that one system will come to interact with, and affect, other systems in surprising ways. These interactions are called *emergent properties* of systems. Another, and somewhat more loaded, term currently in favor is *unintended consequences*. The telephone system not only changed the way people communicated, it changed the nature of communication. Everyone predicted that the automobile would result in people traveling farther and faster, but the modern suburb was an emergent property. Environmental control systems have all sorts of unforeseen consequences on people's health (it worsens), as well as on cities in perpetually hot and humid climates (they're more livable).

Emergent properties regularly affect security systems. Early banks had vaults with simple locks, either key or combination. One early bank robbery technique was to threaten the manager's family until he opened the vault door. Another was to start shooting his co-workers, one by one. No one imagined that incidences of kidnapping among bank managers' families would rise as bank vaults became increasingly impervious to dynamite. (This is one reason modern bank vaults have time locks; managers simply cannot open the lock, even under duress, so threatening them and their families has no purpose.)

In fact, in some ways all security breaches are a result of emergent properties. Locks are supposed to keep people without keys out; lock picking is an emergent property of the system. Even though these properties are at first glance undesirable, their discovery and exploitation by attackers are often the first inkling that the countermeasure is fallible. Security is not a function that can be tested—like a chemical reaction or a manufacturing process. Instead, it is only effectively "tested" when something goes wrong. An insecure system can exist for years before anyone notices its insecurity. Just because your home hasn't been broken into in decades doesn't mean that it's secure. It might mean simply that no one has ever tried to break in, or it might mean that dozens have tried to break in and, without your ever knowing about it, given up in failure. Both of these situations look exactly the same. Because so often a successful security system looks as if nothing ever happens, the only reliable way to measure security is to examine how it fails—in the context of the assets and functionality it is protecting.

••••

At a basic level, security systems are different from any other type of system. Most systems—cars, telephones, government bureaucracies—are useful for what they do. Security systems are useful precisely for what they don't allow to be done. Most engineering involves making systems work. Security engineering involves making sure systems don't fail. It involves figuring out how systems fail and then preventing those failures. As a security expert, I do care about how a system works, but I care more about how it doesn't work. I care about how it reacts when it fails. And I care about how it can be *made* to fail.

Hollywood likes to portray stagecoach robberies as dramatic acts of derring-do: The robbers gallop alongside the fleeing horses and jump onto the careening coach. Actually, it didn't happen that way. Stagecoaches were ponderous vehicles that couldn't negotiate steep hills without everyone getting out and pushing. All the robbers had to do was find a steep hill and wait. When the defensive systems—staying inside the coach, being able to gallop away—failed, the robbers attacked.

Sometimes systems fail in surprising ways. ATMs have all sorts of countermeasures to ensure that a criminal doesn't get hold of customers' account numbers and PINs, because that knowledge can be used to make counterfeit cards that can be used to steal money. In 1993, some enterprising criminals installed a fake ATM at a shopping mall in Manchester, Connecticut. It was programmed to collect the account number from the card and the PIN as it was typed. Once it had harvested this information from an unsuspecting user, the bogus machine would indicate some kind of error or an innocent-sounding "Out of Cash" message, sending cardholders on their way. The criminals left the machine in the mall for a week, then came back and took it away for "repairs." Then they made counterfeit cards with the captured data and used them at real ATMs to steal money.

Designing systems with failure in mind isn't completely foreign to engineers. Safety and reliability engineers also go to great lengths to ensure performance in the face of failure, but there is an important difference between what they do and what security engineers do. When you're designing for safety and reliability, you're designing for a world where random faults occur. You're designing a bridge that will not collapse if there's an earthquake, bed sheets that won't burst into flames if there's a fire, computer systems that will still work—or at least fail gracefully—if the power fails. Sometimes you're designing for large-scale faults—tornadoes, earthquakes, and other natural disasters—and sometimes you're designing for individual faults: someone slipping on the bathroom floor, a child accidentally pushing a button, a tree falling on a building's roof. Security systems need to work under such random circumstances, but they also have to give special consideration to nonrandom events; that is, to the presence of an intelligent and malicious adversary who forces faults at precisely the most opportune time and in precisely the most opportune way.

It is the notion of an adversary that separates safety and security engineering. In security, intelligent opposition is trying to make security fail. And a safety failure caused by an attacker becomes a security failure.

- *Safety*: You can predict how many fire stations a town needs to handle all the random fires that are likely to break out. *Security*: A pyromaniac can deliberately set off more fire alarms than the town's fire stations can handle so as to make his attacks more effective.
- *Safety*: Knives are accidentally left in airplane carry-on luggage and can be spotted by airport X-ray machines. *Security*: An attacker tries to sneak through a knife made of a material hard to detect with an X-ray machine, and then deliberately positions it in her luggage to make it even harder to detect with the X-ray machine.
- *Safety*: Building engineers calculate how many fire doors are required for safe evacuation in an emergency. *Security*: Those doors are deliberately barricaded before murderers set fire to the building. (This happened in a Rwandan convent in 1994.)

A few years ago, a colleague of mine was showing off his company's network security operations center. He was confident his team could respond to any computer intrusion. "What happens if the hacker calls a bomb threat in to this building before attacking your network?" I asked. He hadn't thought of that. The problem is, attackers *do* think of these things. The adolescent murderers at the Westside Middle School in Jonesboro, Arkansas, in 1998 set off the fire alarm and killed five and wounded ten others as they all funneled outside.

Attackers deliberately try to make security systems fail or exploit other people's accidents, and when they find a vulnerability, they can exploit it again and again. Protecting against random faults is different from protecting against malicious attacks. In order to think about security, you need to think about how systems fail.

••••

Even though security is a binary system—either it works or it doesn't—it doesn't necessarily fail in its entirety or all at once. An attacker targets a component of the security system, one piece of the functionality, and tries to use that small failure to break the entire system. A burglar might break into a house by smashing a window, picking the lock on a door, or opening the attached garage and entering the house through the unlocked door there. In each case, the burglar is targeting a particular countermeasure to break the home's security. Sometimes individual failures cascade to ever larger failures. On 11 September 2001, failures in airport passenger screening allowed nineteen hijackers to board four airplanes. Then further security failures aboard the airplanes allowed them to be hijacked. Crashing the planes into the two World Trade Center towers and the Pentagon was the next security failure, although I can't imagine how it could have been defended against at that point. Only in the fourth plane did the last security failure not occur, because those passengers had information that those on the other planes did not.

To rob a bank, burglars might first disable the alarm system, then break into the bank, then break into the vault, and then break into the safe-deposit boxes. The old poem begins: "For the want of a nail, the shoe was lost. For the want of the shoe, the horse was lost. For the want of the horse, the rider was lost. . . ." If you can think about security systems in terms of how individual failures affect the whole, you'll have gone a long way to understanding how security works.

Usually, the easiest way to attack a system isn't head-on. Money in a modern bank vault is secure; money in an armored car is also secure. Money being carried from the armored car on the street into the bank branch and over to the vault is vulnerable. It's increasingly difficult to sneak weapons into an airport through the passenger terminal and the metal detectors, but what about over the fence at night or in the maintenance trucks that regularly go in and out? It's easier to get a fake ID by bribing a suitable government employee than it is to create a high-quality forgery.

Security usually fails at the seams—at the points where two systems interact—seams between security systems and other systems, seams between parts of a security system. Clever attacks focus on the interactions between systems: the side doors forgotten about because everyone has focused on the frontal attacks, the five minutes when the guard is walking around the other side of the building, the shift change at the bank. (High-security prisons do nothing by the clock, and the "changing of the guard" scenes at royal palaces are only elaborate tourist attractions.) Some drug dealers sell only during shift changes at the local police station. Guerrilla warfare has been so effective partly because it specifically targets seams in a modern army's defenses.

Systems have seams in installation. You might have a strong lock on your front door, but did the locksmith who installed the lock save a copy of the key for himself—or for his cousin the felon? Systems have seams in maintenance: The repairman who fixes broken voting machines is in an ideal position to modify their operation. Every storeowner knows that the cleaning service is in an excellent position to clean out the stock.

Sometimes security designers get it right. The Jeweler's Building in Chicago, built in 1926, was designed so that delivery vehicles could drive directly into the building, onto large elevators, and directly to the office floors. This design meant that the weak seam in jewelry transport—moving the goods from the (secure) car into the (secure) building—was also secured.

....

Security systems can fail in two completely different ways. The first way is that they can fail in the face of an attack. The door lock fails to keep the burglar out, the airport face-scanner fails to identify the terrorist, or the car alarm is bypassed by a thief. These are *passive failures*: The system fails to take action when it should. A security system can also fail by doing what it's supposed to do, but at the wrong time. The door lock successfully keeps the legitimate homeowner out, the airport face-scanner incorrectly identifies an honest citizen as a terrorist, or the car alarm rings when no one is trying to steal the car. These are *active failures*: The system fails by taking action when it shouldn't.

Most security systems can fail both ways. A stone wall surrounding your property can have a passive failure—it can fail to be high enough or strong enough—or it can have an active failure—it can collapse unexpectedly on an innocent bystander. An armed sky marshal can either fail to stop a hijacking or can himself terrorize a planeload of passengers in a false alarm.

In most security systems, active failures are more frequent than passive failures. Security countermeasures continually affect the normal functionality of the system, while they only occasionally affect attackers. (Actual attacks are, after all, relatively rare.) This magnifies the effects of active failures, and the impact security systems have on the innocent. Systems with passive failures are simply less effective: They only occasionally succeed in the face of an attack. Systems with a relatively high rate of active failures are almost always more trouble than they're worth because of the high rate of false alarms. And the rarity of real attacks further amplifies the phenomenon.

Imagine an airplane security system that automatically shoots terrorists. I don't know how it works; it's some kind of new magic technology. Even if it had a 50-percent passive failure rate, it might be useful. Shooting half of all the terrorists is a good thing, even if the system misses the other half. But if its active failure rate is just one tenth of a percent, then one in a thousand times it would shoot an innocent passenger, and we would never allow it to be fielded. Assuming that one person in a million is a terrorist, the system would shoot one thousand innocents for every terrorist. That rate of failure is simply unacceptable.

Admittedly, that's an extreme example. Systems that randomly kill innocents are anathema to most people. But many theoretically interesting security systems are useless in practice because of active failures. Car alarms are less effective security devices because they have so many false alarms and many people just ignore them.

A system's active failures not only affect how successful it is, but how likely it is to be implemented in the first place. A bank can't install an ATM that prevents even a small percentage of its customers from getting to their money, no matter how effective it is at keeping thieves from stealing the money. People will not tolerate a door lock that bars them from entering their own home periodically, no matter how effective it is at stopping burglars.

The problem of rare attacks and too many false alarms is important in many security decisions: arming pilots, establishing massive data-mining security systems, and so on. It is far more likely that these systems will react to false alarms or fail by accident than that they will be used to foil an attacker. This high active failure rate makes the countermeasure much less useful in practice and makes it much less likely that the trade-offs necessary to enact the system will be worth it.

It's a perfectly understandable part of human nature to begin to ignore frequent false alarms, but it's also an aspect of human nature that can be exploited by those who attack systems. Like the villagers in the "Boy Who Cried Wolf" story, users will quickly learn to ignore a system that exhibits too many false alarms, with failure as a result. Some versions of Microsoft Windows crash so often that users don't find anything suspicious when a virus, worm, or Trojan induces similar behavior.

Attackers commonly force active failures specifically to cause a larger system to fail. Burglars cut an alarm wire at a warehouse and then retreat a safe distance. The police arrive and find nothing, decide that it's an active failure, and tell the warehouse owner to deal with it in the morning. Then, after the police leave, the burglars reappear and steal everything. Closer to home, it's a common auto theft technique to set a car alarm off at 2 A.M., 2:10, 2:20, 2:30 ... until the owner turns the alarm off to placate angry neighbors. In the morning, the car is gone. It doesn't matter how well the car alarm worked. It didn't make the car secure because the greater system—the owner responding to the alarm—failed.

In Afghanistan in the 1980s, during the war against the Soviets, the rebels would approach a Soviet military base in the middle of the night and throw a live rabbit over the fence. The lights would go on, the alarms would sound, and the Soviets would start firing. Eventually they would figure out that they had killed a rabbit. The following night, the attackers would pull the same trick. After a few nights, the Soviets would turn off the sensors and the rebels would be able to easily penetrate the base.

A similar trick worked against the Soviet embassy in Washington, DC. Every time there was a thunderstorm, a U.S. operative fired sugar pellets against an embassy window. The motion sensors sounded the alarm, which the Soviets would turn off. After a few thunderstorms, the embassy staff disabled the sensors, allowing the attackers to do whatever they had planned.

Systems may be especially vulnerable when rare conditions occur. In the event of a museum fire, who is making sure no one steals the artwork? If everyone is outside watching the prime minister driving by in a motorcade, who is paying attention to a store's cash register? In preindustrial England, public hangings were prime work time for pickpockets, as weddings and funerals still are for burglars. In 1985, days after a volcanic eruption buried the town of Armero in Colombia, a pair of residents came back with picks and shovels and proceeded to rob the bank.

Rarity is also often a cause of active failures. A friend installed a burglar alarm system in his home. The alarm was wired into the burglar alarm company's switchboard; when it went off, the company was automatically alerted, and then it would call the police. My friend had a secret code that he could use to call the alarm company and register a false alarm (the police didn't want to send a squad car out whenever

someone accidentally tripped the alarm and caused a active failure).
He also had a second secret code, a duress code, which meant: There is
a gun being held to my head, and I am being forced to call you and
claim that this is a false alarm. It isn't. Help!

One day my friend accidentally tripped the burglar alarm, and he
dutifully called the alarm company. Inadvertently, he gave them the
duress code instead of the false alarm code. Almost immediately he
realized his mistake and corrected it. The woman from the alarm company gave a huge sigh of relief and said: "Thank God. I had no idea
what I was supposed to do."

When a security event occurs regularly, people become experienced
and know what to do. If the event happens only once every few years,
there could be an entire office staff that has never seen it. Staff members might have no idea what to do. They might ignore anomalies,
thinking they are just system problems and not security incidents.

This is why we all went through fire drills in primary school. We
practiced failure conditions, both so we would be prepared for what
happened and as a constant reminder that the failure could occur. It's
the same with airplanes. When oxygen masks drop from the ceiling,
we don't want the passengers glancing up from their novels, wondering what those silly things are, and then going back to their books.
Nor do we want a nuclear power plant operator wondering what a
flashing red light means because he's never seen it before. Both training and practice are essential because real attacks are so rare.

Poorly designed systems are unprepared for failure. In June 1996,
the European Space Agency's Ariane 5 rocket exploded less than a
minute after launch because of a software error. The possibility of that
failure was never explored because "the error condition could not
happen." In the fifth volume of *The Hitchhiker's Guide to the Galaxy*
series, Douglas Adams wrote, "The major difference between a thing
that might go wrong and a thing that cannot possibly go wrong is that
when a thing that cannot possibly go wrong goes wrong it usually
turns out to be impossible to get at or repair." He has a point.

An important security precept is to expect both passive and active
failures. No matter how well designed a system is, it *will* fail. This fact
has been proven again and again throughout history, and there's no
reason to believe it will ever change. "Unbreakable," "absolute,"
"unforgeable," and "impenetrable" are all words that make no sense
when discussing security. If you hear them, you can be sure you're listening to someone who doesn't understand security or is trying to

hoodwink you. Good security systems are designed in anticipation of possible failure. For example, road signs and highway lights are built with special bases that break away if a car hits them, reducing the damage to both the car and the sign or light.

Cars themselves are designed to fail well upon impact. Before 1959, conventional wisdom was that the way to design a safe car was to make it strong. However, an engineer at Mercedes, Béla Berényi, showed that if the car is too strong, the passengers absorb the force of a collision. He invented something called a crumple zone: a part of the car made of special materials that would crumple in predictable ways, absorbing the energy of a collision and sparing the passengers. This design—crumple zones around a rigid passenger cab—is standard today in every car. By figuring out how cars failed on impact, Berényi was able to design them to fail better, increasing passenger safety.

The *Titanic* was believed to be unsinkable, so there was no reason to have lifeboats for all the passengers. The Enigma machine was unbreakable, so the Germans never believed that the British were reading their encrypted messages. Remember these examples the next time you hear someone talk about a security system that cannot fail.

Chapter 5

Knowing the Attackers

What sort of security is sensible depends, in part, on the types of attackers you're defending against. A burglar, a journalist, or a government agent—each requires a different approach to security. Will the attackers be skilled professionals? How well funded are they? How averse to risk? Will they be motivated by money, or an ideology, or even a personal grudge? Are they outside burglars, employees, or friends and relatives? None of these questions can be answered with certainty, but by taking rational steps to predict your most likely attackers you'll be less likely to waste time and money on unnecessary or inappropriate security.

Security's job is to defend assets, which presupposes that someone wants to attack. Understanding what motivates an attacker and how he might attack those assets is terribly important. This understanding allows you to evaluate the risks he poses and the countermeasures that might thwart him. In *The Godfather, Part II*, Michael Corleone said: "There are many things my father taught me here in this room. He taught me: Keep your friends close, but your enemies closer."

A security alarm that might adequately protect a house is probably ineffective on a military base. Information that is secure against a criminal because it isn't worth any money to her might be insecure against a journalist whose work is being subsidized by a newspaper. Countermeasures that might make you secure against a rational mugger might not be effective against a sniper out for random blood. In a sense, the attacker is a part of a security system, and the system must take that attacker into account to remain secure.

Attackers can be categorized along several basic lines, including their motivations, objectives, expertise, access, resources, and risk aversion. Is the attacker motivated by greed or anger? Is the attacker skilled or unskilled? Funded or unfunded? What is his aversion to his own risk? How ruthless or persistent is he? Is the attacker a lone person or a group? Step 2 of the five-step process asks: What are the risks to the assets you're defending? Understanding your attacker is integral to answering this question.

Take, for example, risk aversion—a measure of what the attacker is willing to lose. Consider what an attacker might risk: perhaps only his reputation (if being caught results in unwanted publicity or public humiliation). Then again, he might risk his job (by getting fired), money (if he is fined or sued), or even his freedom (if he is jailed). In extreme cases, an attacker might risk his life (in very dangerous attacks, or in attacks that involve capital offenses), or he may actively plan to kill himself (in suicide attacks). Most people are risk-averse, of course, but there are notable exceptions. Homeless, penniless drug addicts, for example, can be desperate criminals, because their assessment of risk is so different from that of other people. In the face of their need for a fix, what do they have to lose? Also, people who had something and then lost it are often less risk-averse than those who have much to lose; Karl Marx made this observation in the context of class warfare.

Of course it is not fair to think of only the poor as not being risk-averse. Many others feel they have nothing to lose: the politically disenfranchised, the dispossessed, the aggrieved of all stripes and shades. And the very rich, who can afford to take risks, and teenagers, who know no better, and ... the list can go on and on. The best way to look at potential attackers is to make an effort of the imagination—as perilous as that may be—and to see matters from their perspective. To begin with, you might ask what it is that motivates the particular kind of attacker you think you're facing. For a criminal, it might be money. For an emotional attacker, it might be revenge. For a terrorist, it might be publicity or rewards in the afterlife or the redress of some age-old grievance. A system that doesn't take attackers' personal goals into account is much less likely to be secure against them.

Attackers will, consciously or otherwise, operate within their particular budget constraints and bring to bear only what expertise, access, manpower, and time they have available to them or can obtain elsewhere. Some attacks require access but no expertise: looting a store, for example. Some attacks require expertise but no direct access: hacking into a bank's computer system. Some attacks require some access and some expertise: a car bomb in front of an embassy, for example. All adversaries are going to have a set of attacks that is affordable to them and a set of attacks that isn't. If they are paying attention, they will choose the attack with the maximum benefit for a given cost, or choose the least costly attack that gives them a particular benefit. Basically, they will make a business decision.

Wealthy adversaries are the most flexible because they can trade their resources for other things. They can gain access by paying off an insider, and expertise by buying technology or hiring experts (perhaps telling them the truth, perhaps hiring them under false pretenses). They can also trade money for risk by executing a less risky but more sophisticated—and therefore more expensive—attack. Think of intelligence organizations that use expensive satellite surveillance rather than risking agents on the ground.

If you mischaracterize your attackers, you're likely to misallocate your defenses. You're likely to worry about nonexistent risks and ignore real ones. Doing so isn't necessarily a disaster, but it is certainly more likely to result in one.

••••

A criminal is someone who breaks a law, any law, but there are many different kinds of criminals. Professional criminals (or, at least, avocational criminals) are motivated by profit of some kind: financial, political, religious, social, and the like. If a particular attack isn't profitable, criminals are likely to find something else to do with their time. It's Willie Sutton robbing banks because "That's where the money is," or the Vikings pillaging medieval monasteries.

That motivation leads criminals to attack a variety of assets for a variety of objectives. Most of the time, it's money. Sometimes it's information that could lead to money: the route an armored car will take through the city streets, information about someone's personal life that can be used for blackmail or fraud, and so on. And those objectives lead to a variety of attacks, among them theft, kidnapping, and extortion.

In general, crime doesn't pay. This is why most criminals are stupid or desperate or both; if you're bright enough to calculate the costs and benefits of crime and if you have choices, you're likely to choose another career path. Even so, there are some very intelligent criminals out there. And in places like Russia, where the police are so ineffectual that crime often does pay, smarter people are attracted to crime as a "career."

Different criminals are willing to tolerate different levels of risk: Some are willing to risk jail time but probably don't want to die by sacrificing themselves to the higher calling of armed bank robbery. There are criminals with little expertise and few resources. There are criminals with genuine expertise—master cutpurses or safecrackers, for example—and minimal resources. There are criminals with substantial resources but minimal expertise. And there are organized crime syndicates with both expertise and resources.

Criminals tend not to have good access, with the exception of those who are *insiders*. In fact, really successful crimes generally involve an insider. In the early days of railroads, the insiders were railroad employees who knew the trains' timetables and delivery schedules. They could tell the rest of the gang which trains carried cash and when they would be passing through. Today it might be someone who works inside a bank, an insurance company, or a casino or, better still, someone who is able to manipulate a computer system.

Insiders can be dangerous and insidious attackers. They may have a high level of access to the system they are attacking. Aldrich Ames, who sold government secrets to the Soviets from 1985 to 1994, was in a perfect position within the CIA to sell to the KGB the names of U.S. operatives living in Eastern Europe because he was trusted with their names. Think of a bank employee purposefully mis-setting the time lock to give his burglar friends easier access. Or a stockbroker using proprietary knowledge to manipulate fortunes in the market. Or a casino blackjack dealer palming chips into her own pocket as she deals hands. (Dealers' uniforms no longer have pockets for this reason.) Insiders can be exceedingly difficult to stop because an organization is forced to trust them.

Here's a canonical insider attack. In 1978, Stanley Mark Rifkin was a consultant at the Security Pacific National Bank in Los Angeles. He used his insider knowledge of the money transfer system to move several million dollars into a Swiss account, and then to convert that money into diamonds. He also programmed the computer system to automatically erase the backup tapes that contained evidence of his crime. (He would have gotten away with the theft, except that he bragged about it.)

Insiders perpetrate most of the crime at casinos. Most theft in retail stores is due to those who work in the stores. Disgruntled employees who have badges allowing them access to corporate buildings are a much greater risk than random outside attackers. Insiders are invariably more worrisome attackers than outsiders. Yet perhaps the most common security mistake of all is to expend considerable effort combating outsiders while ignoring the insider threat.

Insiders are not necessarily employees. They can be consultants, suppliers, or contractors, including security guards or cleaning staff: people who remain in a building when everyone else goes home. They may be family members of employees who can get access because of their relationship. Some insiders subvert a system for their own ends: a

McDonald's worker in Monroe, Louisiana, sold drugs from the drive-through window, and a convicted child rapist working in a Boston-area hospital stole a co-worker's password, perused confidential patient records, and made obscene phone calls. Attacks like these are egregious, but probably the most damaging insiders are those who collude with outsiders, augmenting their insider access with other criminal skills. Sometimes insiders can be tricked into inadvertently helping the real attackers.

Many security countermeasures—walls, guards, locks—defend against external attackers but are useless or nearly so against insiders. A king might sleep in a locked bedroom high in his keep, with castle walls, guards, and a moat to protect him, but none of that will prevent his trusted manservant from poisoning his food. Indira Gandhi was assassinated by her own guards. Daniel Ellsberg could release the documents that came to be known as the Pentagon Papers because he had access to them. Workers who clean offices at night have keys to all the doors. While insiders might be less likely to attack a system than outsiders are, systems are far more vulnerable to them.

Insiders know how a given system works and where the weak points are. They may know the organizational structure, and how any investigation against their actions would be conducted. They are trusted to some degree by the system they are going to attack (and the degree of trust they are granted may be extraordinarily high, as was the case with Robert Hanssen, an FBI agent who spied for the Soviets). Insiders can use a system's own resources against itself. They have considerable expertise in the systems they are attacking, and may have been involved in the design or operation of the very systems they attack. Insiders are in a perfect position to commit fraud or sabotage. And even worse, management may be unwilling to admit initially that an attack has come from inside (which is, in effect, an admission of a management failure).

Revenge, financial gain, or institutional change (including institutional destruction) can motivate insiders. Other attackers sometimes can become—or buy—insiders in order to gain knowledge and access: Thieves, industrial competitors, and national intelligence agents have all made extensive use of insiders. Malicious insiders can have a risk aversion ranging from low to high, depending on whether they are motivated by a "higher purpose" or simple greed. Some insiders are professional criminals, and others are just opportunists.

....

Attackers can also be categorized based on their motivations. Some criminals are mere *opportunists*: They didn't lead a life of crime. Instead, they saw an opportunity and took advantage of it. Their characteristics are similar to those of "professional" criminals, except that they're generally more risk-averse. People who drive faster than the speed limit don't want to go to jail; they don't even want the hassle of being stopped for ten minutes by a patrol car. The reason they speed is they're convinced they're not going to get caught.

Think of someone who happens to find an unlocked bicycle or a wallet on the ground. Imagine a casino worker who, for the first time in her life, sees an inch-thick stack of $100 bills. In general, opportunists are much easier to secure systems against, because even cursory countermeasures can dissuade them, and they can be prone to attack without carefully developing a plan or thinking through the consequences. On the other hand, there are far more opportunists than professional criminals. Circumstances can have an effect on even the most honest people. In many countries, it is illegal for the police to deliberately try to trick opportunists into committing crime—it's called entrapment—precisely because of this.

An *emotional attacker* is another type of criminal, one not in it for the money. Think of a person who goes after someone in a bar fight, kids who spray-paint a wall, a gang on a killing spree, or someone killing his lover in a jealous rage. Those in this category include the Columbine kids, disgruntled workers with firearms, abusive husbands, suicide bombers, and those in the throes of road rage. Also included are most computer hackers, hoodlums, and graffiti artists—who commit their crimes because "It's fun." In the U.S., most firearm homicides are crimes of passion. The common thread among these attackers is that their internal reality doesn't square with what most of us take to be objective reality—their attacks are *statement* attacks—and their statements mostly make no sense to others.

Emotional attackers share many characteristics with professional criminals, but their emotionalism often causes them to accept greater risk. The statements they make can range from dramatizing a strike and hurting a retail business by blocking shoppers from entering a market, to blowing a market crowded with shoppers to bloody bits in order to terrorize an entire country. Complicated psychology is at work here; some emotional attackers kill people and then *hope* the cops come to get them—so-called "suicide by cop." Emotional attackers are, as a rule, very hard to defend against. You can't simply say: "No one would commit that attack; it's simply not worth it." To an emotional attacker, it might be. The right set of emotions can lead to any level of risk tolerance.

Friends and relations comprise another category of attackers. They are, in a sense, a special class of insiders. Sometimes their attacks are seriously criminal—a significant amount of credit card fraud is performed by the card owner's relatives—but most of the time these attackers are opportunists. Often it's no more than letting nosiness get the better of them, and they invade a spouse's or friend's privacy by looking through diaries, correspondence, computer files, and wallets. They may be searching for proof of a surprise birthday party or a secret infatuation, siphoning money out of a joint account, or seeking information for a child custody fight.

At the opposite end of the emotional scale are industrial competitors, attackers with the cold intellectual motivation of profit achieved by gaining a business advantage. These are not necessarily criminals; their actions might be perfectly legal—depending on a country's laws—even though they are unwanted by the defending company. In the late 1980s, for example, the German company Henkel built a textile chemical plant in China. By the time it started up, a duplicate plant owned by locals was already in operation down the road. (In China, notions of intellectual property are notoriously different from ours.) An industrial criminal might attempt to steal competitors' trade secrets or sabotage a competitor's plans, or simply be after information—a kind of overreaching way of staying abreast of the competition.

In a widely reported 1993 case, the automotive executive José Ignacio López de Arriortúa was accused of stealing boxes of confidential documents, including automobile and assembly plant designs, when he left General Motors to work for Volkswagen AG. Volkswagen settled the civil case by paying GM $100 million and agreeing to buy an additional $1 billion in GM parts. In an ongoing clash of the titans, Boeing and Airbus have made cross-accusations of espionage for years. The practice of industrial espionage is widespread: Companies from China, France, Russia, Israel, and the U.S., to mention just a few, have all stolen technology secrets from foreign competitors, sometimes with the aid of their governments.

Industrial espionage can be well funded: An amoral but otherwise rational company will devote enough resources to industrial espionage to achieve an acceptable return on investment. Stealing a rival's technology for $500,000 rather than developing it yourself for $5 million can be an enticing trade-off. The overall goal, again, is a business advantage, but remember that a company is not a single monolithic entity and that individual employees will have different agendas. For example, an employee might be more willing to take a risk if he thinks he'll be fired if he doesn't beat the competition.

And not all institutional thefts of information or breaches of corporate privacy involve a company attacking a competitor. The media—the press, wire services, radio, and television—are, of course, very interested in finding out what companies have done and what they may be planning. In essence, all muckraking investigative journalists perform industrial spying but with different motivations and objectives. The press isn't interested in a competitive advantage over its targets; it is interested in "newsworthy" stories and competitive advantages over the rest of the press. Examples include the *Washington City Paper* publishing the video rental records of Judge Bork (which led to the Video Privacy Protection Act of 1988), the *Cincinnati Enquirer* breaking into the voice mail system of a Chiquita executive, or James Bamford publishing a tell-all book about the National Security Agency (NSA). Some reporters have said that they would not think twice about publishing national security secrets; they believe the public's right to know comes first.

In most modern, media-savvy societies, of course, a rich personal scandal involving a public official can give newspaper sales and even television news ratings a substantial boost. At the height of her fame, even marginally compromising photographs of Princess Diana were worth over half a million dollars to circulation-mad tabloids.

This means that the press can bring considerable resources to bear on attacking a particular system or target. Well-funded media outlets have considerable resources; they can hire experts and gain access. And if they believe the information they're uncovering is important, they can tolerate a surprising amount of risk. (Certainly Bob Woodward and Carl Bernstein, the reporters who broke the story of the 1972 Watergate break-ins, fall into this category.) Reporters in the U.S. and in other countries have gone to jail to protect the identity of their sources and their right to publish what they believe is true. Some have died trying to get a story. The press can be an example of an attacker on the side of good that most of us support.

· · · ·

Any list of potential attackers must include lawful governments, particularly police departments. This is so for several reasons:

- Against criminals, the police are attackers, and any study of security needs to consider that fact dispassionately.
- The police are not always on the side of good. The problem with rogue law enforcement is especially acute in countries where the police are the enforcement arm of a corrupt regime. In such places, clearly, we *want* to teach people how to defend themselves against the police. In Russia, for example, the police are poorly paid and regularly work for rich businessmen who employ them as enforcers. In North Korea, law enforcers must obey the particular agenda of Kim Jong Il's government. In many Latin American countries, the police are more like organized crime, extorting money from the local populace and independent criminals alike. The poorly paid Mexican police, for example, routinely make extra money by issuing traffic tickets and keeping the proceeds.
- Even the most just societies have rogue policemen. J. Edgar Hoover used the FBI as a private enforcement tool. In recent years in the U.S., we can find examples of police officers planting evidence, giving false testimony, using illegal interrogation techniques, being bribed to work for criminals, and just plain terrorizing the populace they're entrusted with defending. These are exceptions, of course, but they are common enough to worry about.

Given their special powers of access, their legal rights to bear arms, and their special powers in terms of interrogation, arrest, and detention, the police absolutely must be analyzed in terms of their ability to act as attackers.

As attackers, the police have a reasonable amount of funding and expertise. They're moderately risk-averse. No police officer wants to die, but they routinely risk their lives to save others—and especially their fellow officers. And since they have the law on their side, many attacks are less risky to the police. For example, a search warrant turns the crime of breaking and entering into a lawful investigative tool.

In the U.S. and many other countries, citizens have deliberately put in place all sorts of laws that hamper or constrain the police—for example, limits on lawful interrogation, search and seizure rules, and rules for gathering evidence. These constraints have been created not to make the police ineffective, but because people know that the inevitable police abuses would be worse were the laws not in place. Countries without these sorts of laws are invariably less safe places to live.

Just as the police can be considered an attacker, so can national intelligence organizations. All of them are dedicated and capable, with a world of expertise and access, and are extremely well funded. U.S. intelligence budget figures are classified, but are said to hover between $30 billion and $35 billion in recent years—a figure likely to have ballooned in the post-9/11 world.

While these organizations can be extremely formidable attackers, their actions are not without limitations. To begin with, intelligence organizations are highly risk- and publicity-averse. Movies portray spies as dashing agents who engage in high-speed car chases while defusing a bomb and dodging shoulder-fired missiles, but a huge amount of what goes on in these organizations is mind-bogglingly boring. They don't like to see their names on the front page of *The New York Times* and generally don't engage in risky activities (except, of course, the failures that we read about in the *Times*). Exposed operations cause two problems. First, they reveal what data has been collected. Secret information is the basis of national intelligence. Whether it's eavesdropping on a negotiating position, sneaking a peek at a new weapons system, knowing more than the adversary does—all such information is much more valuable if it's secret. If the adversary learns what the intelligence organization knows, some of the benefit of that knowledge is lost.

The second, and probably more important, problem is that botched operations, and especially botched publicized operations, expose techniques, capabilities, and sources. For many years, the NSA eavesdropped on Soviet car phones as politburo members drove around Moscow. Then someone leaked information about Khrushchev's health to the newspapers, and soon thereafter all the politburo car phones were outfitted with encryption devices. The newspapers didn't print anything about car phones, but the KGB was able to connect the dots. The important leak wasn't that we knew about Khrushchev's health, but that we were listening to their car phone communications. The same thing happened after the terrorist bombing of the La Belle Berlin discotheque in 1986. President Reagan announced that we had proof of Libya's involvement, compromising the fact that we were able to eavesdrop on Libyan embassy traffic (messages) to and from Tripoli. One of the major reasons the

U.S. was reluctant to produce its evidence of Iraq's weapons programs in 2002 and 2003 was the risk of compromising sources and methods.

Leaks and scandals aside, national intelligence agencies are as a rule very effective because they have the full weight of a government behind them, and they are constantly working to develop and deploy a wide variety of methods and tactics. When Israeli security questions you at the Tel Aviv airport, one of the things they may do is take your laptop computer into another room and return it to you some minutes later. My guess is that they copy selected contents of the hard drive for examination later, without any probable cause or warrant.

Finally, as far as attackers go, a military at war is about as formidable an attacker as you can get. The military has more expertise and resources than any other adversary. Its members are highly motivated and can tolerate great risks.

Some countries have extremely skilled and well-funded militaries, while other countries have timid and passive militaries. Democratic countries can be risk-averse, although it depends very strongly on circumstance. Autocracies like Saddam Hussein's Iraq, Kim Jong Il's North Korea, and Nazi Germany, and theocracies like Khomeini's Iran and the Taliban's Afghanistan, are willing to engage in behavior far riskier than that of most other countries on the planet.

Military force constitutes an attack, but a discussion of it and of military defense is, for the most part, beyond the scope of this book. If an army is invading your country, please look elsewhere.

. . . .

The word *terrorist* is a catchall term for a broad range of ideological groups and individuals, both domestic and international, who use displays of violence for political goals. Terrorist groups are usually motivated by geopolitics or ethnoreligion—Hezbollah, IRA, ETA, Red Brigade, Shining Path, Tamil Tigers, FLNC, PKK, KLA—but can also be motivated by moral and ethical beliefs, like Earth First and radical antiabortion or animal rights groups. Terrorism is not a movement or an ideology, but a military tactic. A terrorist is someone who employs physical or psychological violence against noncombatants in an attempt to coerce, control, or simply change a political situation by causing terror in the general populace. The desire to influence an audience is an essential component of terrorism. Someone who goes on a crime spree, whether a deranged serial killer who chooses victims in alphabetical order or a sniper who chooses victims at random, is not a terrorist. Someone who launches a guerrilla campaign against a military is not a terrorist (at least for the purposes of this book). Japanese kamikaze pilots and Iraqi paramilitary bombers are not terrorists. Lucas Helder put a series of pipe bombs in mailboxes around the Midwest in an effort to leave a pattern of explosions in the shape of a smiley face on the map—craziness, certainly, but not terrorism. Someone who straps a bomb filled with poison-soaked nails to his waist and blows himself up in a crowded restaurant to protest the ruling government—that's a terrorist.

The primary aim of a terrorist is to make a statement; news of the attacks is more important than the attacks themselves. Without widespread publicity, the attacks lose much of their effectiveness. Marshall McLuhan wrote: "Terrorism is an ingenious invention by which any two or more armed people can take over an entire billion dollar industry with the complete cooperation, not only of its workers, but of its owners." The rise of global media and the omnipresence of television news only magnifies the effect. Would the 9/11 terrorists have had nearly the same psychological impact had the video clip of the second plane hitting the south tower not been played over and over again on newscasts globally? Probably not. Today's media make attacks more effective and tempting because they give publicity-seeking attackers much more reach and thus much more leverage. Terrorists can be assured of almost instant worldwide coverage of their acts and dissemination of their position. And the attacks terrorists gravitate toward—bombings and shootings—have the visceral visual impact the media love.

Terrorists are not going to be deterred by security systems that make attacks unprofitable in the conventional sense. Terrorist attacks aren't even going to be deterred by security systems that deter normal emotional attackers, because terrorists measure success differently. The best way to deter terrorist attacks is to deny terrorists their goal by giving them only minimal media coverage. Admittedly, doing this is difficult in practice.

The characteristics of terrorists vary. They can be well funded and have considerable resources, or they can operate on a shoestring. They can have considerable skills, or they can have none. (Building a modern bomb takes skill, but a crude device is easy to make. Of course, both can be effective.) They often have good access to their targets. And, of course, terrorists are willing to tolerate an enormous amount of risk, in some cases including certain self-destruction.

Terrorists often have their own rationality. While the rest of the world might view terrorist actions as completely irrational, that doesn't mean the terrorists have the same view. It might be obvious to you that blowing up a bus of schoolchildren is not the best way to get a seat at the bargaining table, but historically terrorism often has led to an overall climate of futility and despair and eventually negotiation. The unfortunate truth is that in modern society, terrorism often does work.

Terrorists can do things that are blatantly counter to the survival of the species and the planet. For a long time it was believed that terrorists would never unleash a deadly plague because it would affect their side just as much as it would affect their enemies. But imagine a terrorist thinking: "I'm in the Third World. We don't have much of anything. We have an agrarian economy. If we lost a third of our population, our culture would survive. Our economy would go on as before, albeit with fewer people. But the U.S. is a rich industrial nation. If they lost a third of their population, they would be utterly devastated. Maybe a deadly plague would level the playing field." With that kind of analysis, terrorists are capable of compassing some truly cataclysmic events. In trying to understand a terrorist's motivations, sometimes it is more useful to look at who has more to lose, not who has more to gain.

Many terrorists are willing to die during an attack. These days, some of them *plan* to die. This isn't new—there are lots of historical examples of suicidal terrorists—but it's certainly more in the news these days. Suicide bombers are best thought of as part of the machine used to execute the attack. Instead of spending billions of dollars researching and designing autonomous robots that could take over and pilot an airplane, Al Qaeda used suicidal people.

Terrorists often consider themselves at war with their enemies, but they don't represent traditional military powers or nation-states. This fact makes traditional threats of military retaliation less effective against them than they would be against a nation or government. The invasion of Afghanistan certainly helped disrupt Al Qaeda, but there simply isn't a way to retaliate as there would be if, for example, a government were behind the attacks.

Chapter 6

Attackers Never Change Their Tunes, Just Their Instruments

A defender must consider both likely attacks and likely attackers. Attacks have not varied very much throughout history: Murder is murder, stealing is stealing, and fraud is fraud. But even as the fundamental goals and motivations of attackers remain constant, the tools and techniques of attack change. A single attacker can undertake many different attacks, and a single type of attack can usually be launched by a number of different attackers. Each of these attacker–attack combinations constitutes a threat. All of the threats, taken together and weighted for both their likelihood and the seriousness of their consequences, collectively determine the risks. Critical to evaluating a security system is to determine how well it mitigates risk.

There hasn't been a new crime invented in millennia. Murder, theft, impersonation, counterfeiting—they are all perennial problems and have been around since the beginning of society. The ancient Egyptians wrote about terrorism in their lands. The Code of Hammurabi, circa 1780 B.C., provides penalties for various forms of theft. Around 390 B.C., Xenophon wrote that the Spartans deliberately raised boys in a way that would lead them to steal. The Romans complained about counterfeiting, corruption, and election fraud.

Even biological attacks aren't new. Around 600 B.C., the Athenian lawgiver Solon fouled the water supply of Krissa, a city he was attacking, with hellebore root. The defenders came down with severe cases of diarrhea. In medieval Europe, it was a common offensive tactic to catapult dead animals into besieged cities in an attempt to foment disease. The Tartars launched plague-stricken corpses into Kaffa in 1346. During the French and Indian War, the British deliberately gave Indian tribes smallpox-infected blankets. And in 1797, Napoleon tried to infect residents of the besieged city of Mantua, Italy, with swamp fever.

The types of attacks are timeless because the motivations and objectives of attackers are timeless. What does change is the nature of attacks: the tools, the methods, and the results. Bank robbery is a different crime in a world of computers and bits than it is in a world of paper money and coinage. Identity theft is today's name for the crime of impersonation.

Fraud has been perpetrated against every commerce system ever invented. Unscrupulous merchants have rigged scales to shortchange their customers; people have shaved silver and gold off the rims of coins. Everything has been counterfeited: currency, stock certificates, credit cards, checks, letters of credit, purchase orders, casino chips. Modern financial systems—checks, credit cards, and ATM networks: each racks up multimillion-dollar fraud losses per year.

Fraud has also been perpetrated against individuals for all of recorded history. Occasionally new confidence scams are invented—Ponzi schemes are probably the newest form of fraud—but the basic mechanisms are ancient.

One category of fraud worth singling out is what I call *I'm Sorry* attacks. These are easy-to-make errors that attackers can make intentionally, then simply apologize for if caught. They're attacks without penalty, because in these attacks it's almost impossible to prove malicious intent. Shortchanging customers is of course the easiest example. If the attacker is spotted, he apologizes, blushes, and fixes the problem. If not, he pockets the difference. These attacks aren't always small-scale; the 1980s term for this kind of thing when done by a government was *plausible deniability*. President Reagan's entire defense when his government was caught selling weapons to Nicaragua in violation of the law was basically: "I didn't know it was happening."

The Internet has made some other kinds of frauds easier. Successful scams that used to be sophisticated undertakings—requiring office space, phone numbers, letterheads—now only require a good-looking Web site. In August 2000, Mark Jakob sent a forged press release, purporting to be from Emulex Corporation, to a newswire; the fake news sent Emulex stock plummeting 62 percent. Jakob, who'd taken a short position on the stock, made money when the share price declined. In October 2002, West African criminals used a fake version of a British bank's online service to defraud unsuspecting customers by stealing account information and passwords. The same technology that allows a small company to compete with a multinational allows a scam Web site to compete with a legitimate one.

Identity theft is often portrayed as a fraud of the modern era, but crimes involving impersonation are much older. Martin Guerre was a prosperous peasant who lived in Artigat, France, near Toulouse. In 1548, when he was twenty-four years old and married, he disappeared. Eight years later someone who seemed to be Martin returned, greeting people by name and recalling details about them and his past interactions with them. His wife greeted him as her husband. He moved back into his home, had two more children, and life went on.

But there was talk that the Martin who had returned was an imposter named Arnaud du Tilh. In 1560, there was a trial. More than thirty townspeople swore that he was the real Martin Guerre; more than forty swore he wasn't. The court ruled that he was an imposter and sentenced him to death. He appealed to the Parliament of Toulouse. On the day the Parliament was to deliver its verdict—reports are that he was to be acquitted—a one-legged man hobbled in and claimed to be the real Martin Guerre. The court questioned them both, found the defendant guilty again, and hanged him.

Modern identity theft is more about electronic impersonation than physical impersonation. And it's much more often a tool of a simple thief who wants to use someone else's checks or credit and debit cards, as well as take advantage of the person's past good credit history. One Albuquerque, New Mexico, criminal ring would break into homes specifically to collect checkbooks, credit card statements, receipts, and other financial mail, looking for Social Security numbers, dates of birth, places of work, and account numbers: information that could be used for identity theft. Other criminals are collecting identity information off the Internet; online genealogical databases are helpful to find someone's mother's maiden name. The goal is to take this information, open bank accounts and obtain credit cards in the person's name, get large cash advances, and then disappear, leaving the victim with the debt. In 2001, some say there were over 500,000 cases of identity theft in the U.S. alone. (I've seen even higher estimates.) There are even "Hints from Heloise" columns on protecting yourself.

Identity theft is one of the fastest-growing areas of fraud because it has recently become a relatively safe, and very profitable, crime. Electronic identity, which is becoming more and more common, is increasingly easy to steal. There are many ways to steal someone's identity—bank statements and other documents in the trash, database files on the Internet—and there are many credit card companies competing for new business. (How many of us have document shredders at home? How many of us have crosscut shredders?) At the same time, identity theft is becoming more profitable. What used to require a face-to-face visit can now be done by mail, phone, or the Internet.

．．．．

Assaults that range in severity from graffiti, to arson, to military bombing campaigns fall into the general category of destructive attacks. These attacks do not necessarily target tangibles; intangibles can be destroyed, as well. (Think of attacks against someone's good name, or against a company's brand or reputation.) Banks don't often publicize when their computers have been hacked and money has been stolen because the loss of customer confidence, if the attack were made public, is worth much more than the money stolen. When Tylenol capsules were poisoned, individual consumers suffered a destructive attack against their person, and Johnson & Johnson suffered a destructive attack against its brand and good name.

Modern technology facilitates long-distance destructive attacks. Military tactics might come to mind first—warplanes stationed in the U.S., and even commanded from the U.S., involved in bombings on the other side of the globe. But the practice of long-distance attacks has civilian implementations, as well. In 1994, Citibank's computers were attacked by criminals from St. Petersburg, Russia, who stole millions of dollars. More recently, individual U.S. computer users have been defrauded by criminals operating in Eastern European and Asian countries.

Suicidal attackers form their own class of destructive attackers. Suicide attacks are an old idea. Japan has an ancient tradition of honorable suicide, for example, and during World War II, the Japanese used the notion of suicidal pilots flying planes into targets to great effect. Kamikazes were very difficult to defend against. Ships were sunk, and thousands of Allied seamen died. And that was against hardened targets that knew the attacks were coming.

．．．．

Then there are attacks that are not destructive at all—at least not immediately. Consider attacks on privacy. In 1890, Samuel Warren and Louis Brandeis (who would later become a Supreme Court Justice) wrote an influential article defining privacy as "the right to be left alone." Even so, the U.S. has a mixed notion of privacy: Homes are sacrosanct, but personal data can be bought and sold freely. European countries tend to have fewer laws protecting physical privacy and more protecting data privacy, although safe harbor agreements with the U.S. have watered those down considerably. Asian countries like Singapore have less privacy protection overall. Lots of people think they have a right to privacy; in fact, that right depends on the laws of the country they live in.

And there may even be times you don't want privacy. You might, for example, appreciate it when Amazon.com suggests books and music you might like, based on your buying habits and the buying habits of others with similar tastes. But it's a two-edged sword: If a bank knew that I was an indifferent comparison shopper, would it offer me a higher interest rate the next time I needed a loan? (Is it illegal for the bank to do so? Should it be?) Certainly a car dealership would like to know beforehand which of its customers are not good hagglers.

One sort of privacy violation that is little discussed is traffic analysis: information about communication patterns, as opposed to information about the contents of communications. Consider the following example: Even if the secret police can't listen in on your telephone conversations, phone records showing whom you call and how long you talk are available in the U.S. without a warrant, and they are of enormous value. The Nazis used phone records to round up associates of people who'd already been arrested. In the weeks after 9/11, American authorities used this kind of information to investigate people who'd phoned the terrorists. National intelligence organizations make use of this kind of information all the time, to keep tabs on who is involved with whom. Assuming it is turned on, your cell phone also provides a pretty accurate history of where you've been—on some networks down to a several-block radius—which the police can use in an investigation (and to find you in case of emergency). And recently, the Colombian drug cartels have started using phone records to figure out who in their group are phoning government officials—and then executing them as informants.

Along similar lines, simply watching the number of cars in the Pentagon parking lot after hours on different days, or the number of late-night pizza deliveries, can tell you when a military initiative is imminent. Even worse, detailed pictures of the Pentagon's parking lots can be bought from several commercial satellite companies.

Privacy invasions can be the first step in a greater crime. Identity thieves first invade privacy to get the information they need to impersonate people. Many stalkers have been able to track their victims from information in databases. The canonical example is Hollywood actress Rebecca Schaeffer, killed in 1989 by a stalker who got her home address from public DMV records. Schaeffer's death resulted in a federal law requiring the DMV to keep home addresses confidential. Less gruesome but just as effective was a ring of thieves in Iowa who collected license plate numbers off luxury cars in an airport's long-term parking lot, retrieved home addresses from DMV substations, and then robbed the unoccupied homes.

These examples demonstrate one of the primary dangers of data aggregation. In creating large databases of information, whether government or corporate, we are opening ourselves to the possibility that the databases will be subverted by attackers. Regardless of the value of large government databases, such as that proposed as part of the 2002 Total Information Awareness program, the far more likely outcome is that the information will be used by attackers—outsiders breaking in, rogue government employees, corporations with good lobbyists—to invade the privacy of citizens whose information is being collected.

Maybe this kind of thing is inevitable. Can we even expect any privacy in today's technological world? More and more, mechanisms to facilitate surveillance are built into our society's infrastructure: the Internet, the cell-phone network, the proliferation of public and private security cameras. Some people, like science fiction author David Brin, have advocated a "transparent society," in which security is enhanced precisely because everybody can see everything. Technology has turned the battle for and against privacy into an arms race: The side with the more advanced (i.e., expensive) technology wins, with a large advantage going to those who would violate privacy. Laws, to be sure, can step in where technology fails; privacy violations can be made illegal. This may work well—but only against attackers who obey the law.

. . . .

Attackers can execute many different attacks, and the same attack can be executed by many different types of attackers. An attacker plus a series of potential attacks, all with the same goal, is a threat. A thief stealing a car by smashing the window or jimmying the lock is a threat. Your sister sneaking into your room, via the door or the window, and reading your diary, is a threat.

We need to understand who the attacker is so we can determine which security countermeasures will work and which ones won't. For example, if something is digging up your garden, you may need to set a trap—but the kind of trap will differ depending on whether you suspect the culprit is a rabbit, the neighbor's dog, or the kid down the street.

Similarly, we need to understand what attacks might be undertaken in order to defend against them. Since an attacker might use many possible attacks to achieve the same outcome, understanding the attacker's desired outcome is essential to defending against him.

For example, a terrorist who wants to create havoc will not be deterred by airline security; he will simply switch to another attack and bomb a shopping mall. However, a countermeasure that results in the terrorist getting arrested is effective in stopping whatever attack he might have planned.

Of course, the possible attacks depend on the scope of the assets you are defending, which in turn depends on your agenda. Effective security countermeasures will be different if you are trying to defend an aircraft, an airline, the commercial air-transport system, or the nation from terrorism. If you're an airline and your agenda is defending airplanes from terrorism, a countermeasure that protects one plane but allows a terrorist to blow up another is ineffective, but a countermeasure that protects all your planes and forces the terrorists to attack shopping malls is effective. If you're a homeowner and your agenda is defending your home from burglary, a countermeasure that protects your house but allows the burglar to rob the house across the street is perfectly effective; if you're the local police force, it isn't.

Understanding the assets you're protecting in terms of a system is vital. Not only must you look at the assets plus the functionality around those assets, you also must properly define those assets.

The decongestant pseudoephedrine used to be available for purchase in bottles of 100 tablets. Manufacturers assumed the main security risk regarding this drug was shoplifting, and they let stores deal with the problem. But when the manufacturers of methamphetamine (speed) started buying the pills for use as a raw material, the U.S. government began to regulate their sale in large bottles. In response, the manufacturers stopped selling large bottles and changed the packaging so that each individual tablet had to be punched out of a cardboard, paper, and plastic pouch by hand ... not worth the trouble for drug makers. Thus, a new attacker appeared with a new attack, and a completely new countermeasure had to be devised to deal with the resultant new threat.

The goal of the attacker matters a great deal. There is an enormous difference between an attacker who is looking for any victim and an attacker who is targeting a specific one. An old joke ends with the line: "I don't have to outrun the bear; I just have to outrun you." This is true only if the bear doesn't care whom it catches, as long as it catches someone. But if I were the only one covered in honey, that would be something else again. (Remember, though, that in real life bears generally chase people to scare them, not to eat them.)

Quite a few security systems—home security, car security, and some personal security—are designed to make the attacker go somewhere else. The operative assumption in these systems is that the burglar doesn't really care which house in the neighborhood he robs; any one will do. If you have an alarm system and your neighbor doesn't, the burglar is likely to go next door. Even though he might be able to defeat your burglar alarm, why should he bother? On the other hand, if you have a priceless painting hanging in your living room and the burglar knows it, he's not going to go next door. It's your alarm system he'll try to defeat.

The more prominent you are—person, organization, or government—the more likely you are to attract targeted attacks, especially if you're suddenly thrust into the limelight. If you're a company, for example, expect targeted attacks to increase after a major layoff, a stock-price drop, or an Enron-like scandal. This is also true on a local scale: If it's widely known that you have the best computer or stereo in the entire dorm, you're at a greater risk of theft.

What this means is that you need to judge a risk based not only on who is likely to attack the system and what they want, but on exactly what assets are being attacked. Maybe the terrorists don't care what aircraft they hijack, or maybe they specifically want to hijack an El Al flight. Maybe the kidnapper targets a category of people—businessmen or women—but doesn't care about the specific people. A fired employee doesn't want revenge on just any company; she wants revenge on the company that fired her. And someone angry at the post office isn't going to be satisfied shooting up the Motor Vehicles Bureau; he may even want revenge against a specific post office.

Attackers invent new attacks. It's a feedback mechanism in the system: Smart attackers react to the countermeasures and modify their attacks accordingly. Remember that scene in *Raiders of the Lost Ark* when one of the bad guys does a very fancy set of scimitar moves and Indiana Jones just takes out a gun and shoots him? The audience chuckled at that because Indiana did the unexpected. He cheated. Attackers regularly cheat. Like the 9/11 hijackers, they act in unanticipated ways. In Chapter 5, I mentioned that the Vikings attacked monasteries because that's where the money was. The reason treasures were kept in monasteries was because the Christian God protected them. No one would loot a monastery, out of fear of retribution in the afterlife. That countermeasure didn't work on the Vikings; they weren't Christian, so they didn't believe in afterlife retribution.

On 24 November 1971, someone with the alias Dan Cooper invented a new way to hijack an aircraft, or at least a new way of getting away. He took over a Northwest Orient flight from Portland to Seattle by claiming he had a bomb. On the ground in Seattle, he exchanged the passengers and flight attendants for $200,000 and four parachutes. Taking off again, he told the pilots to fly at 10,000 feet toward Nevada. Then, somewhere over southwest Washington, he lowered the plane's back stairs and parachuted away. He was never caught, and the FBI still doesn't know who he is or whether he survived. (Rumors abound, though.) The attack exploited a vulnerability in the seams of the security system: We spend a lot of effort securing entry and exit to aircraft on the ground but don't really think about securing it in the air. (Also notice the cleverness in asking for four parachutes. The FBI had to assume that he would force some of the pilots and remaining crew to jump with him and could not risk giving him dud chutes.) Cooper cheated and got away with it.

In the 1990s, a group from MIT cheated casinos out of millions at blackjack. While casinos have sophisticated systems for detecting card counters, they are all designed to find individuals. The MIT group used teams of individuals, none of whom did anything alone to arouse suspicion, but all of them together were able to tip the odds in their favor. The unexpected often defeats existing countermeasures.

Burglars cheat, too. Homes are generally protected by door and window locks, and sometimes by alarm systems. Modern "Hole in the Wall Gangs"—in Las Vegas, Miami, New York, and undoubtedly elsewhere—break through the walls of the homes they rob. By reframing the problem, they render most home-security countermeasures irrelevant. In the natural world, some attackers use similar methods. Some flowers that are pollinated effectively by hummingbirds have long, tube-like shapes to try to prevent all their nectar from being taken by short-tongued nectar drinkers, like bees, which wouldn't pollinate them as effectively. It's a simple and elegant security solution: The hummingbirds, with their long bills and long tongues, easily get the nectar, but the short-tongued freeloaders simply can't reach. Except that some bees just chew a hole in the side of the flower and get the nectar that way.

(It's no accident that insects provide such interesting security examples. Over the eons they have tried just about everything. The techniques of attack and defense that proved to work were repeated, and the techniques that failed weren't. Because they tried them at random and stopped at the first workable solution they found, they tended to arrive at interesting and surprising solutions. It's not unusual to find insect countermeasures that are nonobvious, but effective nonetheless. Insects are good at cheating.)

Cooper's airplane hijacking serves to illustrate another point. After his story hit the press, there was an epidemic of copycat attacks. In thirty-one hijackings the following year, half of the hijackers demanded parachutes, and the technique even came to have its own name—parajacking. It got so bad that the FAA required Boeing to install a special latch—the Cooper Vane—on the back staircases of its 727s so that they couldn't be lowered in the air. (Notice the pattern: A vulnerability is discovered and then repeatedly exploited until eliminated.)

Most attackers are copycats. They aren't clever enough to react to feedback from the countermeasures and invent new techniques. They can't think up new ways to steal money out of ATMs or bank vaults. They don't have the vision to deploy their infantry divisions in anything but the same formations they've seen repeatedly in the past. They attempt the same attacks again and again, or they read stories of attacks in the newspaper and think: "Hey, I could do that, too." As long as the old attacks still prove to be effective, the attackers who use them are a dangerous adversary.

Once you know the threats, the next step is to look at their effects and figure out the risks. The possible attacks determine the threats; the likely, or devastating, attacks determine the risks. Some minor threats are minor risks (the threat of space aliens invading Earth), and some major threats are major risks (the threat of credit card fraud.) Some minor threats are major risks even though they're remote, because their effects are so significant (the threat of terrorists detonating a nuclear device), and some major threats are minor risks because they're too insignificant to worry about (the threat of a mosquito flying into the house).

Examining an asset and trying to imagine all the possible threats against that asset is sometimes called "threat analysis" or "risk analysis." (The terms are not well defined in the security business, and they tend to be used interchangeably.) Some consultants charge a great deal of money to perform these analyses for companies and to reach what they consider comprehensive and definitive conclusions. But frankly, most of the time we don't have enough information to accurately calculate risks.

Threat and risk analyses are hard to do well, and performing them is a skill that comes only with effort and experience. Analysis involves

figuring out how a system can be attacked and how the security countermeasures can be bypassed. It involves trying to put yourself in the
mind of an attacker. The best security analysts naturally go through
life looking for ways to break the security of systems. They can't walk
into a polling place without figuring out ways that they can vote twice.
They can't use a credit card without thinking about the possible
antifraud mechanisms and how to get around them. They look for
blind spots in stores' video surveillance systems.

Try it yourself. The next time you dine in a restaurant, think of all
the ways you can eat a meal and then not pay. (Can you get a free meal
by faking the business card of a local restaurant reviewer?) Think of all
the ways you can be attacked while eating. (Did you ever notice that
when you leave your car to be valet-parked, the attendant has both
your address—on the registration in the glove compartment—and
your garage door opener? Even so, it's a risky attack for him because
he would be high on any list of suspects.) It's fun, and the whole table
can participate. Then try to figure out which threats are serious risks,
which are middling risks, and which are not worth worrying about.
This is harder, but the real goal.

The point of this exercise is to determine a system's risks. Once
you do that, you can seek out and implement security countermeasures
to mitigate those risks. You can never reduce the risks to zero, but you
can reduce them significantly if you are willing to make whatever
other trade-offs are required.

Threat and risk analysis must apply to the system as a whole. To
think about risk properly, you can't just focus on the risks of a burglar
picking a lock on the store's door or disabling the alarm system. You
need to take a holistic approach: Who is attacking the system? What
do they want? What are the characteristics of their attack? How likely
is it that they will attack? Security may fail at the component level;
only a systemic view of security can see the entire problem.

And remember, attackers can use many different attacks to
achieve their goal. If an airport mitigates the risk of someone sneaking
a bomb through passenger security, it also has to worry about the
threat of someone sneaking a bomb in via a maintenance truck. If the
airport mitigates that risk, there will be something else to worry about.
It makes more sense to mitigate both risks partially than it does to
mitigate one risk completely and ignore the other one.

Example: Credit Card Numbers on the Internet

Should you avoid typing your credit card number onto a Web page or sending it to a merchant via e-mail in order to prevent credit card fraud? Let's take it through our five steps.

Step 1: What assets are you trying to protect? You're trying to protect your credit card number.

Step 2: What are the risks to these assets? You're trying to solve the problem of credit card theft or, at least, to make sure the problem doesn't happen to you. Hackers and other criminals steal credit card numbers on the Internet. You want to avoid their stealing yours. You do want to keep in mind, though, that it's less of a risk than it could be. In the U.S., the maximum loss a customer can suffer is $50, often waived in the event of fraud. The worst part of the ordeal is the hassle of getting a new card and updating your many automatic payment systems with the new number.

Step 3: How well does the security solution mitigate those risks? The solution you're considering is not sending your card number over the Internet. However, hackers steal credit card numbers by the tens of thousands in merchant databases that are connected to the Internet. You probably have, in your wallet right now, a credit card whose number has been stolen. As a customer, your card number gets into those databases regardless of whether you place your order by mail, by phone, or over the Internet. Hackers aren't going to waste time stealing your number individually. You're at risk once you make a purchase using the credit card; the method you use doesn't really matter.

Step 4: What other risks does the security solution cause? None that I can think of.

Step 5: What trade-offs does the security solution require? Making purchases over the Internet is convenient, easy, and fast. It can also get us the best price. Not being able to do this would be a significant adverse trade-off.

Given all this, is the security solution—not using your credit card on the Internet—worth it? No. The countermeasure described in Step 3 does not mitigate the risk described in Step 2. Given that, there's no point in implementing the countermeasure.

Example: Security Screening at Airports

After 9/11, the government implemented a tough new screening policy at airports, having security personnel confiscate even the smallest knives and almost anything else that looked sharp or pointy. Besides employing metal detectors and X-ray machines, the agents randomly culled people in line for further screening using a metal-detecting hand wand, and manually searched their bags.

Many of us suspected that this didn't do any good, but let's see if we can figure out why.

Step 1: What assets are you trying to protect? The answer to this question depends on who you are. If you're a passenger, you're concerned about protecting the flight you're on. If you're an airline, you're concerned about protecting all your airplanes. If you're the FAA, you're concerned about protecting air traffic in general. If you're the government or a citizen, you're concerned about protecting the people from terrorism, and airplanes are only a small piece of the assets you're trying to protect.

Step 2: What are the risks to these assets? The obvious answer is: aircraft terrorism. A more exact answer is: weapons in the hands of passengers while they are on the plane. The assumption is that keeping weapons out of the hands of potential terrorists on airplanes will help stop terrorism on board an aircraft. It will stop some—stupid terrorists who try to bring a gun, a machete, or a homemade bomb on an airplane with them.

Step 3: How well does the security solution mitigate those risks? Only marginally well. Even enhanced screening doesn't identify all weapons. For example, in tests between November 2001 and February 2002, screeners missed 70 percent of knives, 30 percent of guns, and 60 percent of (fake) bombs. I'd expect the statistics to be even worse in nontest situations; real attackers would angle weapons in the carry-on luggage, or take them apart so they'd be even harder to spot. We can't keep weapons out of prisons, a much more restrictive and controlled environment. How can we have a hope of keeping them off airplanes?

Additionally, as has been pointed out again and again in essays on the ludicrousness of post-9/11 airport security, anything can be a weapon: a rock, a battery for a laptop computer, a belt, or the bare hands of someone with the proper training. You can buy a composite knife that doesn't show up on a metal detector. Snap the extension handle of a wheeled suitcase off in just the right way, and you've got a pretty effective spear. Break a bottle and you've got a nasty weapon. Garrotes can be made of fishing line or dental floss. I remember having a screwdriver confiscated at a checkpoint in the Newark, New Jersey, airport and buying a replacement at a store inside the security perimeter. Two months later I wandered around Washington National Airport, inside security, and found for sale in the airport shops all the ingredients I needed to build an incendiary device.

You can even make a knife onboard the plane. Buy some steel epoxy glue at a local hardware store. It comes in two tubes: a base with steel dust and a hardener. Make a knifelike mold by folding a piece of cardboard in half. Then mix equal parts from each tube and form into a knife shape, using a metal fork from your first-class dinner service (or a metal spoon you carry aboard) for the handle. Fifteen minutes later you've got a reasonably sharp, very pointy, black steel knife. All of this is to illustrate that security screening will never be effective at keeping all weapons off aircraft. On the other hand, even a partially effective solution may be worthwhile.

Step 4: What other risks does the security solution cause? Quite a few, in fact. The screeners themselves have to be trusted, since they can introduce weapons into the secure area. The screening technology has to be trusted, because if it fails, security isn't being provided. There are security risks inherent in bunching up people in lines before the security checkpoints. And the general irritation of passengers is another, albeit small, security problem; it might contribute to air rage.

Step 5: What trade-offs does the security solution require? Security screening is expensive, and the public will have to pay for it one way or another. Perhaps the airlines will pay for it and then pass the cost along to consumers by raising ticket prices. Or perhaps a separate tax will be added to the ticket price. Or perhaps the government will pay for it and raise taxes accordingly. However you look at it, this manpower-intensive system will cost.

It's also time-consuming. Lines have gotten much shorter since the first months after 9/11, but security still takes longer than it used to. And it's intrusive. Many people have complained about their luggage being opened and searched in public. There have been all sorts of stories about insensitive baggage screeners and embarrassed fliers, and more than a few lawsuits. And it can be dangerous; for at least a year after 9/11, diabetics could not take necessary needles aboard planes.

This analysis tells us that, to a point, airport screening is worth it. However, after a single cursory screening step, the system rapidly reaches the point of diminishing returns. Most criminals are stupid; they're going to try to smuggle a weapon on board an aircraft, and there's a decent chance that a screening system will stop them. Just as important, the very act of screening every passenger in an airport acts as a reminder and a deterrent. Terrorists can't guarantee that they will be able to slip a weapon through screening, so they probably won't try. But we must face the fact that a smart criminal will figure out a way around any screening system, even a laborious and intrusive one. The way to prevent airplane terrorism is not to keep objects that could fall into the wrong hands off of airplanes; a better goal is to keep those with the wrong hands from boarding airplanes in the first place.

As the scope of your security concerns in Step 1 broadens, the countermeasure becomes less and less effective because the assets are larger and more varied. Preventing terrorism on airplanes is particularly important because planes have catastrophic failure properties: If something goes wrong, everyone on the plane dies. But countermeasures that simply move terrorist attacks from airplanes to shopping malls are less effective than countermeasures that neutralize terrorists. Mitigating the risk is much harder if the risk is terrorism in general.

Chapter 7

Technology Creates Security Imbalances

Throughout history, technological innovations have changed the power balance between attacker and defender. Technology can give the advantage to one party or the other, and new attacks may be possible against technologically advanced systems that were not possible against older, simpler systems. Concepts such as automation, class breaks, and action at a distance also increase threats by giving attackers more leverage; and the sheer complexity of modern systems makes them more vulnerable. Understanding how advances in technology affect security—for better or for worse—is important to building secure systems that stand the test of time.

When dynamite was invented in 1886, European anarchists were elated. Finally they had a technology that they believed would let them dismantle the state . . . literally. It didn't work out that way, of course; European anarchists found that blowing up a building wasn't as easy as it seemed. But shaped charges and modern explosives brought down the U.S. Marine and French military barracks in Beirut in October 1983, killing 301.

Throughout history, technological innovations have altered the balance between attacker and defender. At the height of its empire, Rome's armies never lost a set battle, precisely because their equipment and training were far better than their opponents'. The stirrup revolutionized warfare by allowing people to fight effectively on horseback, culminating in the formidable medieval mounted knight. Then the crossbow came along and demonstrated that mounted knights were vulnerable after all. During the American Civil War, the Union Army used the railroad and telegraph to create a major security imbalance; their transport and communications infrastructure was unlike anything the Confederacy could muster, and unlike anything that had come before.

In the eighteenth century, Tokugawa Japan, Manchu China, and Ottoman Turkey all maintained large armies that were effective for internal security; but having eschewed modern technology, they were highly vulnerable to European powers. The British army used the machine gun to create another security imbalance at the Battle of Ulundi in 1879, killing Zulus at a previously unimaginable rate. In World War I, radio gave those who used it intelligently an enormous advantage over those who didn't. In World War II, the unbalancing technologies were radar, cryptography, and the atomic bomb.

When NORAD's Cheyenne Mountain complex was built in the Colorado Rockies in 1960, it was designed to protect its occupants from nuclear fallout that would poison the air for only a few weeks. Provide a few weeks' sustenance, and Cheyenne Mountain workers would be safe. Later the cobalt bomb was invented, which could poison the air for many years. That changed the balance: Use a cobalt bomb against the mountain, and a few weeks' air supply means nothing.

What has been true on the battlefield is true in day-to-day life, as well. One hundred years ago, you and a friend could walk into an empty field, see no one was nearby, and then have a conversation, confident that it was totally private. Today, technology has made that impossible.

It's important to realize that advanced tools—a technological advantage—do not always trump sheer numbers. Mass production is its own technological advantage. In the city of Mogadishu alone there are a million assault rifles (and only 1.3 million people); they're for sale in the local markets and are surprisingly cheap. And they are relatively easy to use. En masse, these small arms can do just as much damage as a biological attack, or a dozen Apache helicopters. Given all these guns, or a million machetes in Rwanda—mass-produced weapons of mass destruction—even a disorganized mob, if it is large enough, can create devastation.

Nonetheless, smart attackers look for ways to make their attack as effective as possible, and technology can give them more leverage and more ambition. Physically robbing a bank can net you a bagful of money, but breaking into the bank's computer and remotely redirecting bank transfers can be much more lucrative and harder to detect. Stealing someone's credit card can be profitable; stealing her entire identity and getting credit cards issued in her name can be more profitable; and stealing an entire database of credit card numbers and associated personal information can be the most profitable of all.

The advance of technology is why we now worry about weapons of mass destruction. For the first time in history, a single attacker may be able to use technology to kill millions of people. Much has been written about these weapons—biological and nuclear—and how real the threats are. Chemical weapons can kill people horribly, though on a smaller scale. But the point is that if a certain weapon is too cumbersome, or too expensive, or requires too much skill for attackers to use today, it's almost certain that this will be less true a year from now. Technology will continue to alter the balance between attacker and defender, at an ever-increasing pace. And technology will generally favor the attacker, with the defender playing catch-up.

This is not an immediately obvious point, because technological advances can benefit defenders, as well. Attackers know what they're doing and can use technology to their benefit. Defenders must deploy countermeasures to protect assets from a variety of attacks, based on their perception of the risks and trade-offs. Because defenders don't necessarily know what they're defending against, it is more expensive for them to use technological advances at the precise points where they're needed. And the fact that defenders have to deal with many attackers magnifies the problem from the defenders' point of view. Defenders benefit from technology all the time, just not as efficiently and effectively as attackers do.

Another important reason technology generally favors attackers is that the march of technology brings with it an increase in complexity. Throughout history, we've seen more and more complex systems. Think about buildings, or steam engines, or government bureaucracies, or just about anything else: Newer systems are more complex than older ones. Computerization has caused this trend to rocket skyward. In 1970, someone could tune up his VW bug with a set of hand tools and a few parts bought at the store down the street. Today he can't even begin the process without a diagnostic computer and some very complex (and expensive) equipment.

Computers are more complex than anything else we commonly use, and they are slowly being embedded into virtually every aspect of our lives. There are hundreds of millions of lines of instructions per computer, hundreds of millions of computers networked together into the Internet, and billions of pieces of data flowing back and forth.

Systems often look simple because their complexity remains hidden. A modern faucet may seem simpler than having to bring water up from the well in buckets, but that's only because there's a complex indoor plumbing system within your house's walls, a mass-production system for precision faucets, and a citywide water distribution system to hook up to. You might only do a few simple things on your computer, but there's still an unimaginable amount of complexity going on inside. When I say that systems are getting more complex, I am talking about systems as a whole, not just the part the average person sees.

As a consumer, I welcome complexity. I have more options and capabilities. I can walk to my bank to withdraw my money, or I can use my ATM card in a cash machine almost anywhere in the world. I can telephone my bank and have them print and mail a check to a third party, or even mail me a money order. Or I can do these things on the Internet. Most people continually demand more features and more flexibility, and more complexity evolves to deliver those results. The consequence is a bounty of features and conveniences not even imagined a decade or two ago.

But of course there's a hitch. As a security professional, I think complexity is terrifying. It leads to more and more subtle vulnerabilities. It leads to catastrophic failures, which are both harder to test for beforehand and harder to diagnose afterward. This has been true since the beginning of technology and is likely to be true for the foreseeable future. As systems continue to get more complex, they will continue to get less secure. This fundamental tension between ever more complex systems and security is at the core of many of today's security problems. Complexity is the worst enemy of security. Computer pioneer Niklaus Wirth once said: "Increasingly, people seem to misinterpret complexity as sophistication, which is baffling—the incomprehensible should cause suspicion rather than admiration."

Complex systems have even more security problems when they are *nonsequential* and *tightly coupled*. Nonsequential systems are ones in which the components don't affect each other in an orderly fashion. Sequential systems are like assembly lines: Each step is affected by the previous step and then in turn affects the next step. Even Rube Goldberg–type machines, as complicated as they are, are sequential; each step follows the previous step. Nonsequential systems are messier, with steps affecting each other in weirdly complex ways. There are all sorts of feedback loops, with machines obeying various do-this-if-that-happens instructions that have been programmed into them. Nonsequential systems are harder to secure because events in one place can ripple through the system and affect things in an entirely different place. Consider the air passenger transport system—a dream, and sometimes a nightmare, of nonsequential design. Each plane affects every other plane in the vicinity, and each vicinity affects still more remote locations. Weather problems in one city can affect take-offs and landings on the other side of the globe. A mechanical failure in Hong Kong, or a wildcat strike in Rome, can affect passengers waiting to board in Los Angeles or New York.

Tightly coupled systems are those in which a change in one component rapidly sets off changes in others. This is problematic because system effects occur quickly and sometimes unpredictably. Tightly coupled systems may or may not be sequential. Think of a row of dominos. Pushing one will topple all the others. The toppling is sequential, but tight coupling means that problems spread faster and that it's harder to contain them.

An AT&T phone crash in 1990 illustrates the point. A single switch in one of AT&T's centers failed. When it came back online, it sent out signals to nearby switches causing them to fail. When they all came back on, they sent out the same signals, and soon 60,000 people were without service. AT&T had a backup system, of course, but it had the same flaw and failed in exactly the same way. The "attacker" was three lines of code in a module that had been recently upgraded. The exacerbating factor was that the test environment was not the same as the real network. Similarly, a computer worm unleashed in Korea in January 2003 affected 911 service in Seattle.

A business, on the other hand, is complex but reasonably sequential and not tightly coupled. Most people work for a single boss. If someone calls in sick, or quits, the business doesn't fall apart. If there's a problem in one area of the business, the rest of the business can still go on. A business can tolerate all sorts of failures before it completely collapses. But it's still not perfect; a failure at the warehouse or the order entry desk could kill the entire business.

．．．．

By changing the functionality of the systems being defended, new technologies can also create new security threats. That is, attacks not possible against the older, simpler systems are suddenly possible against the newer, more complex systems.

In 1996, Willis Robinson worked at a Taco Bell in Maryland. He figured out how to reprogram the computerized cash register so that it would record a $2.99 item internally as a one-cent item while visibly showing the transaction as the proper amount, allowing him to pocket $2.98 each time. He stole $3,600 before he was caught. The store manager assumed a hardware or software error and never suspected theft—remember what happens when attacks are rare—and Robinson got caught only because he bragged. This kind of attack would not have been possible with a manual or even an old electric cash register. An attacker could under-ring a sale and then pocket the difference, but that was a visible act and it had to be repeated every single time $2.98 was stolen. With the computer helping him, Robinson was able to steal all the money at the end of the day no matter who was ringing up purchases.

Before we had the ability to bank by phone, there was no way for an attacker to steal money out of an account remotely. No system was in place for him to attack. Now such a system exists. Before companies started selling on the Internet, it was impossible for an attacker to manipulate the prices on Web pages and convince the software to sell the merchandise at a cheaper price. Now that kind of attack takes place regularly. Manual systems are not vulnerable to attacks against the power system; electrical systems are. Pencil-and-paper systems aren't vulnerable to system crashes; computerized systems are.

Technological systems also require *helpers*: mediators and interfaces between people and the system they interact with. Telephone keypads, aircraft instrument panels, automatic language translation software, and handheld GPS receivers are all helpers. They add even more complexity to a system, and they introduce a level of abstraction that can obscure the system's security properties from the people using it.

Imagine you're sitting on an isolated park bench talking to a friend. That's a relatively simple system to secure against eavesdropping. If you're talking on the telephone, it's a much harder security problem. Not only is there a complex technological system between you and your friend, but there is also a helper device—a telephone— mediating your interaction with both the telephone network and your friend. In this case, you have to secure not only your environment and your friend's environment, but two telephones and the vast and unseen telephone system, as well.

Securing a theater against someone sneaking in and watching the movie is a much simpler problem than securing a pay-per-view television program. Making paper money hard to counterfeit is much easier than securing a credit card commerce system or an electronic funds transfer system. As helpers mediate more and more interactions, the complexity of the systems supporting those interactions increases dramatically. And so do the avenues of attack. It's like the difference between a manual missile launcher and a remote satellite missile launcher. In the former case, the only way to attack the system is to attack the people doing it. In the latter case, you could also attack the satellite system.

In addition, helpers obscure the security-related properties of a system from the users and make it even harder for them to assess risk accurately. It's easy to understand security systems surrounding cash sitting in a wallet. In contrast, the security systems required to support a check-writing system and, even more so, a credit card or a debit card system are much less obvious. By hiding the details, helpers make a system easier to operate at a basic level, but much harder to understand at the more detailed level necessary to evaluate how its security works.

This problem is larger than security. Designers put a good deal of effort, for example, into coming up with familiar metaphors for technological systems. We're all familiar with the trash can icon on computer desktops, e-mail symbols that show sealed envelopes flying out of your out box when you press the "send now" key, and online photo albums. New computer users are regularly surprised when they throw a file away into the trash can and it later turns out to be still in existence on their hard drives, or when they learn that an e-mail message is really much more like a postcard than it is like a sealed letter. Science fiction author Neal Stephenson calls this phenomenon "metaphor shear"—the technological complexities or unpalatable facts that we hide or smooth over in the interests of functionality or easy understanding.

Technological advances bring with them standardization, which also adds to security vulnerabilities, because they make it possible for attackers to carry out *class breaks*: attacks that can break every instance of some feature in a security system.

For decades, phone companies have been fighting against class breaks. In the 1970s, for example, some people discovered that they could emulate a telephone operator's console with a 2600-Hz tone, enabling them to make telephone calls for free, from any telephone.

In the mid-1980s, someone discovered that a Mexican one-peso coin (then worth about half a cent) was indistinguishable from the $1.50 token used for toll machines on the Triborough Bridge and Tunnel Authority's bridges in New York. As word spread, use of the Mexican coins became more common, eventually peaking at 7,300 coins in one month. Around the same time, another clever attacker discovered that Connecticut Turnpike tokens ($0.175 each—they were sold in packages of ten) would work in New York City subway turnstiles ($0.90); the same company made both tokens. (Coin sensors to detect the precise composition of the metal as well as the size and weight, an effective countermeasure, are not cost-effective for such a low-value application. They are more likely to be found on high-value applications like slot machines, where an attacker can win thousands of dollars.) Class breaks can be even worse in the computer world, in which everyone uses the same operating systems, the same network protocols, and the same applications. Attacks are regularly discovered that affect a significant percentage of computers on the Internet.

Picking the lock to a garage requires the attacker do some work. He has to pick the lock manually. Skill acquired picking other garage locks will certainly make the job easier, but each garage he attacks is still a new job. With the invention of automatic garage-door openers, another type of attack became possible. A smart attacker can now build a single garage-opening device that allows him to effortlessly open all the garages in the neighborhood. (Basically, the device cycles through every possible code until the garage door responds.)

Class breaks mean that you can be vulnerable simply because your systems are the same as everyone else's. And once attackers discover a class break, they'll exploit it again and again until the manufacturer fixes the problem (or until technology advances in favor of the defender again).

Automation also exacerbates security vulnerabilities in technological systems. A mainstay of technology, automation has transformed manufacturing (from automobiles to zippers), services, marketing, and just about every aspect of our lives. Automation is also a friend to attackers. Once they figure out a class break, they often can attack a greater number of systems by automating the attack. Automation makes individual attackers, once they've perfected a break, much more dangerous.

Manually counterfeiting nickels is an attack hardly worth worrying about, but add computers to the mix and things are different. Computers excel at dull and repetitive tasks. If our counterfeiters could "mint" a million electronic nickels while they slept, they would wake up $50,000 richer. Suddenly the task would have a lot more appeal. There are incidents from the 1970s of computer criminals stealing fractions of pennies from every interest-bearing account in a bank. Automation makes attacks like this both feasible and serious.

Similarly, if you had a great scam to pick someone's pocket, but it worked only once every 100,000 tries, you'd starve before you successfully robbed anyone. But if the scam works across the Internet, you can set your computer to look for the 1-in-100,000 chance. You'll probably find a couple dozen every day. This is why e-mail spam works.

During World War II, the British regularly read German military communications encrypted with the Enigma machine. Polish mathematicians first discovered flaws in the cryptography, but basically the system was broken because the British were able to automate their attack in ways the German cryptographers never conceived. The British built more than 500 specialized computers—the first computers on the planet—and ran them round-the-clock, every day, breaking message after message until the war ended.

Technology, especially computer technology, makes attacks with a marginal rate of return and a marginal probability of success profitable. Before automation, it was possible for defenders to ignore small threats. With enough technology on the attackers' side, defenders often find that the small stuff quickly gets too big to ignore.

Automation also allows class breaks to propagate quickly because less expertise is required. The first attacker is the smart one; everyone else can blindly follow his instructions. Take cable TV fraud as an example. None of the cable TV companies would care much if someone built a cable receiver in his basement and illicitly watched cable television. Building that device requires time, skill, and some money. Few people could do it. Even if someone built a few and sold them, it wouldn't have great impact.

But what if that person figured out a class break against cable television? And what if the class break required someone to press some buttons on a cable box in a certain sequence to get free cable TV? If that person published those instructions on the Internet, it could increase the number of nonpaying customers by millions and significantly affect the company's profitability.

In the 1970s, the Shah of Iran bought some intaglio currency printing presses and, with the help of the U.S. government, installed them in the Tehran mint. When Ayatollah Khomeini seized power, he realized that it was more profitable to mint $100 bills than Iranian rials. The U.S. Treasury Department calls those bills "supernotes" or "superbills," and they're almost indistinguishable from genuine notes. It got so bad that many European banks stopped accepting U.S. $100 bills. This is one of the principal reasons the U.S. redesigned its currency in the 1990s; the new anticounterfeiting measures employed on those bills are beyond the capabilities of the presses in Iran. But even though there was no way to prevent the supernotes from entering circulation, the damage was limited. The Iranian presses could print only so much money a year, which put a limit on the amount of counterfeit money that could be put into circulation. As damaging as the attack was, it did not affect monetary stability.

Now imagine a world of electronic money, money that moves through computer networks, into electronic wallets and smart cards, and into merchant terminals. A class break against a system like that would be devastating. Instead of someone working in his basement forging individual bills, or even Iranians working in factories forging stacks of bills, a forger could write a computer program that produced counterfeit electronic money and publish it on the Internet. By morning, it could be in the hands of 1,000 first-time counterfeiters; another 100,000 could have it in a week. The U.S. currency system could well collapse as everyone realized that most of the electronic money in circulation was counterfeit. This attack could do unlimited damage to the monetary system.

This kind of thing happens on computer networks all the time; it's how most of the major Internet attacks you read about happen. Smart hackers figure out how to break into the networks, and they write and promulgate automated tools. Then anyone can download the ability to break into computer networks. These users of automated tools are known as script kiddies: clueless teenagers who don't understand how the attacks work but can use them anyway. Only the first attacker needs skill; everyone else can just use the software.

Encapsulated and commoditized expertise expands the capabilities of attackers. Take a class break, automate it, and propagate the break for free, and you've got a recipe for a security disaster. And as our security systems migrate to computerized technology, these sorts of attacks become more feasible. Until recently, we were able to think about security in terms of average attackers. More and more today, we have to make sure that the most skilled attackers on the planet cannot break the system, because otherwise they can write an automatic tool that executes the class break and then distribute that tool to anyone who can use it anywhere in the world.

Another effect of modern technology is to extend reach. Two hundred years ago, the only way to affect things at a distance was with a letter, and only as quickly as the physical letter could be moved. Warfare was profoundly different: A ruler would give his general orders to do battle in a foreign land, and months would go by before he heard word of how his armies did. He couldn't telephone his general for an update.

Today he can. Not only can he instantly learn how his armies are doing, but he can also approve individual bombing targets from the comfort of his office. Airplanes, long-range missiles, remote drones, satellite reconnaissance: These give a country the ability to effect such weighty acts at a distance. Soon we'll have remote-controlled tanks and artillery pieces. Already some aspects of war are becoming an extension of a video game; pilots sitting at an American base can steer UAVs (Unmanned Aerial Vehicles) in remote war zones using a computer screen and a joystick.

Technology facilitates action at a distance, and this fact changes the nature of attacks. Criminals can commit fraud with their telephones. Hackers can break into companies using networked computers. And eventually—not today, for the most part—cyberterrorists will be able to unleash undreamed-of havoc from the safety of their home countries or anywhere else.

Action at a distance makes systems harder to secure. If you're trying to secure your house against burglary, you only have to worry about the group of burglars for whom driving to your house is worth the trouble. If you live in Brussels, it doesn't matter to you how skilled the house burglars are in Buenos Aires. They are not going to get on an airplane to rob your house. But if you run a computer network in Brussels, Argentine attackers can target your computer just as easily as they can target any other computer in the world. Suddenly the list of potential attackers has grown enormously.

The notion of action at a distance also affects prosecution, because much of our criminal justice system depends on attackers getting close to victims and attacking. If attackers are safe within their own country but attack you in your country, it's harder to have them arrested. Different countries have different laws. The student who wrote the ILOVEYOU computer virus in 2000 lived in the Philippines. He did an enormous amount of damage to networks around the world, but there was no Philippine law he could be accused of breaking. One of the primary motivations for the U.S. invading Afghanistan was that the ruling Taliban was not willing to arrest Al Qaeda members.

The international nature of the Internet will continue to be a persistent problem. Differences in laws among various countries can even lead to a high-tech form of jurisdiction shopping: An organized crime syndicate with enough money to launch a large-scale attack against a financial system would do well to find a country with poor computer crime laws, easily bribable police officers, and no extradition treaties. Even in the U.S., confusion abounds. Internet financial fraud, for example, might be investigated by the FBI, the Secret Service, the Justice Department, the Securities and Exchange Commission, the Federal Trade Commission, or—if it's international—the Customs Service. This jurisdictional confusion makes it easier for criminals to slip through the seams.

Data aggregation is another characteristic of technological systems that makes them vulnerable. The concept is old, but computers and networks allow data aggregation at an unprecedented level. Computers, by their nature, generate data in ways that manual and mechanical systems do not. In supermarkets, computerized cash registers attached to bar code product scanners generate a list of every item that every customer purchases, along with the date and time he purchased it. Early mechanical cash registers generated no persistent data at all; later models generated a simple register tape. Online bookstores generate not only a list of everything every customer buys, but everything every customer looks at and how long the customer looks at it. Modern digital cell phones can generate a record of every place the phone is while it is turned on, which presumably is also a record of where the user is, day and night.

This data generation goes hand in hand with new breakthroughs in data storage and processing; today data is easily collected, correlated, used, and abused. If you wanted to know if I used the word "unguent" in a paper copy of this book, you would have to read it cover to cover. If you had an electronic copy of this book, you could simply search for the word. This makes an enormous difference to security. It's one thing to have medical records on paper in some doctor's office, police records in a police station, and credit reports in some bank's desk; it's another thing entirely to have them in electronic form. Paper data, even if it is public, is hard to search and correlate. Computerized data can be searched easily. Networked data can be searched remotely and then collated, cross-referenced, and correlated with other databases. Automated accessibility is one of the emergent properties of the cheapness of data storage combined with the pervasiveness of networks.

Under some circumstances, using this data is illegal. People have been prosecuted for peeking at confidential police or tax files. Many countries have strong privacy laws protecting personal information. Under other circumstances, it's called data mining and is entirely legal.

For example, the big credit database companies have mounds of data about nearly everyone in the U.S. This data is legally collected, collated, and sold. Credit card companies have a mind-boggling amount of information about individuals' spending habits: where they shop, where they eat, what kind of vacations they take. Grocery stores give out frequent-shopper cards, allowing them to collect data about the food-buying proclivities of individual shoppers. Other companies specialize in taking this private data and correlating it with data from other public databases.

And the data can be stolen, often en masse. Databases of thousands (in some cases millions) of credit card numbers are often stolen simultaneously. Criminals break into so-called secure databases and steal sensitive information all the time. The mere act of collecting and storing the data puts the data at risk.

Banks, airlines, catalog companies, and medical insurers are all saving personal information. Many Web sites collect and sell personal data. The costs to collect and store the data are so low that many companies just say, "Why not?" These diverse data archives are moving onto the public networks, where they're not only being bought and sold, but are vulnerable to all sorts of computer attackers. The system of data collection and use is now inexorably linked to the system of computer networks, and the vulnerabilities of the latter system now affect the security of the former.

. . . .

The common theme here is leverage. Technology gives attackers leverage because they can do more in an attack. Class breaks give attackers leverage because they can exploit one vulnerability to attack every system within a class. Automation gives attackers leverage because they can exploit vulnerabilities millions of times. Technique propagation gives attackers leverage because now they can try more attacks, including ones they can't even understand. Action at a distance and aggregation also give attackers leverage because now there are many more potential targets.

Attackers can exploit the commonality of systems among different people and organizations. They can use automation to make unprofitable attacks profitable. They can parlay a minor vulnerability into a major disaster.

Leverage is why many people believe today's world is more dangerous than ever. Thomas Friedman calls this new form of attacker the "superempowered angry young man." A lone lunatic probably caused the anthrax attacks in October 2001. His attack wouldn't have been possible twenty years ago and presumably would have been more devastating if he'd waited another twenty years for biowarfare technology to improve. In military terminology, a leveraged attack by a small group is called an "asymmetric threat." Yale economist Martin Shubik has said that an important way to think about different periods in history is to chart the number of people ten determined men could kill before being stopped. His claim is that the curve didn't vary throughout much of history, but it's risen steeply in the last few decades. Leverage is one of the scariest aspects of modern technology because we can no longer count on the same constraints to limit the effectiveness of attackers.

Thinking about how to take technologies out of the hands of malicious attackers is a real conundrum precisely because many of the most effective tools in the hands of a terrorist are not, on the face of it, weapons at all. The 9/11 terrorists used airplanes as weapons, and credit cards to purchase their tickets. They trained in flight schools, communicated using e-mail and telephone, ate in restaurants, and drove cars. Broad bans on any of these technologies might have made the terrorists' mission more difficult, but it would change the lives of the rest of us in profound ways.

Technology doesn't just affect attackers and attacks, it affects all aspects of society and our lives. It affects countermeasures. Therefore, when we make a technology generally available, we are trading off all of the good uses of that technology against the few bad uses, and we are risking unintended consequences. The beneficial uses of cell phones far outweigh the harmful uses. Criminals can use anonymous e-mail to communicate, but ban it and you lose all the social benefits of anonymity. It's the same with almost every technology: Automobiles, cryptography, computers, and fertilizer are all good for society, but we have to accept that they will occasionally be misused.

This is not to say that some technologies shouldn't be hard to obtain. There are weapons we try to keep out of most people's hands, including assault rifles, grenades, antitank munitions, fighter jets, and nuclear bombs. Some chemicals and biological samples can't be purchased from companies without proper documentation. This is because we have decided that the bad uses of these technologies outweigh the good.

But since most of technology is dual-use, it gives advantages to the defenders, as well. Moonshiners soup up their cars, so the revenuers soup theirs up even more (although these days it's more likely drug smugglers with speedboats in the Caribbean). The bad guys use encrypted radios to hide; the good guys use encrypted radios to coordinate finding them.

And sometimes the imbalances from a new technology naturally favor the defenders. Designed and implemented properly, technological defenses can be cheaper, safer, more reliable, and more consistent—everything we've come to expect from technology. And technology can give the defenders leverage, too. Technology is often a swinging pendulum: in favor of attackers as new attack technologies are developed, then in favor of defenders as effective countermeasures are developed, then back in favor of attackers as even newer attack technologies are developed, and so on.

Technology can also make some attacks less effective. Paper filing systems are vulnerable to theft and fire. An electronic filing system may be vulnerable to all sorts of new computer and network attacks, but the files can be duplicated and backed up a thousand miles away in the wink of an eye, so burning down the office building won't destroy them.

Still, fast-moving technological advances generally favor attackers, leaving defenders to play catch-up. Often the reasons have nothing to do with attackers, but instead with the relentless march of technology. In the 1980s, Motorola produced a secure cell phone for the NSA. Cell phone technology continued to march forward, phones became smaller and smaller, and after a couple of years the NSA cell phone was behind the times and therefore unpopular. After we spent years making sure that telephone banking was secure enough, we had to deal with Internet banking. By the time we get that right, another form of banking will have emerged. We're forever fixing the security problems of these new technologies.

The physicist C.P. Snow once said that technology "… is a queer thing. It brings you great gifts with one hand, and it stabs you in the back with the other." He's certainly right in terms of security.

Chapter 8

Security Is a
Weakest-Link Problem

Attackers are more likely to attack a system at its weak points, which makes knowledge of the weakest link critical. In part this depends on knowing your likely attackers: What's the weakest link for a suicide bomber will be different from the weakest link for a petty burglar. All systems have a weakest link, and there are several general strategies for securing systems despite their vulnerabilities. *Defense in depth* ensures that no single vulnerability can compromise security. *Compartmentalization* ensures that a single vulnerability cannot compromise secure entirely. And *choke points* reduce the number of potential vulnerabilities by allowing the defender to concentrate his defenses. In general, tried and true countermeasures are preferable to innovations, and simpler overlapping countermeasures are preferable to highly complex stand-alone systems. However, because attackers inevitably develop new attacks, reassessment and innovation must be ongoing.

"A chain is no stronger than the weakest link." It's an axiom we've understood since childhood. No matter how strong the strongest links of a chain are, no matter how many strong links there are in it, a chain will break at its weakest link. Improve the strength of the weakest link and you improve the strength of the chain. Whatever you do to any other link of the chain won't make it stronger.

A lot of security is like this. If your house has two doors, the security of the house is the security of the weaker one. If you can identify a dozen different methods to smuggle a gun onto an airplane, the security of the airplane depends on the easiest of the methods. If you're trying to protect a head of state in his office, home, and car, and while he's in public, his overall safety level is no higher than his safety level in the most insecure of those locations. Smart attackers are going to attack a system at the weakest countermeasure they can find, and that's what a security system has to take into account.

Ever had a bag of chips you just couldn't open? It's sealed everywhere, and the material is too resistant to tear. But if you can put the slightest notch in the bag with your teeth, it tears easily at that point. Your attack had two phases: create a weak link and then exploit it.

Military strategists call defense "the position of the interior." Defenders have to consider every possible attack. They need to remain alert the entire time they're guarding. Attackers, on the other hand, have to choose only one attack and can concentrate their forces on that one attack. They decide when to attack, needing to find only one moment when the defenders are not completely alert. This is the reason that bodyguards are overpowered so often: They can't defend against every possible attack at all possible times.

Just as security is subjective, so is the weakest link. If you think you've found the weakest link in a system, think again. Which weakest link did you find? The one for a petty criminal? The one for organized crime? The one for a terrorist? There might be one weakest link for an attacker who is willing to kill, and another one for an attacker who is not.

Finding the weakest link is hard enough, and securing it may not be worth the trade-offs. Imagine a town's inhabitants building a common defensive wall around their homes. Each family is responsible for its own part of the wall, and all the parts connect to form a solid perimeter. Of the hundred families building the wall, ninety-nine of them build it tall and strong. If one slacker family, though, builds a segment that's especially short and weak, they create the weakest link. Much of everyone else's effort has been for naught.

To improve security, the town leaders must raise the lowest part of the wall until it reaches the desired height, find the *new* lowest part of the wall and raise it enough so that *it* is at the desired height, and continue until the remaining lowest part of the wall is at or above the desired security level. Doing this requires a detailed security analysis before starting, and ongoing analysis as things change. But it ensures that security improvements are made exactly where they're most needed. It also means that increasing overall security is expensive, because increasing the security of the wall means raising it at many parts.

Security improvements rarely happen that way, of course. More often than not, we improve security haphazardly, recognizing a problem and fixing it in isolation—often without looking at the whole system, without identifying whether it is the weakest link. In the town wall parable, it's easy to see which part of a wall is the lowest, but complex systems are much harder to understand and analyze. On the other hand, a complex system also makes an attacker's job harder, because she, too, might not be able to find the weakest link.

We can elaborate the parable of the town wall to illustrate some of the ways to minimize the problems caused by the weakest link, all of which we'll talk about in this chapter. The first is called *defense in depth*. Two walls—one behind the other—are often better than one taller wall and work to keep deer out of your garden. And a single wall with razor wire at the top can be a far more effective countermeasure than two walls without it. Compartmentalization is another effective countermeasure: Suppose that instead of one wall around the entire city, the townspeople erected separate walls around each building. Then an attacker who climbed one wall would gain access to only one building and would not be able to attack the entire city. Creating choke points is another technique: If you can force attackers to climb only at a few sections of the wall, you have much less wall to worry about securing.

Defense in depth is often just common sense. It's protecting assets with not one countermeasure, but multiple countermeasures. A good perimeter defense—door locks and window alarms—is more effective combined with motion sensors inside the house. Forgery-resistant credit cards work better when combined with online verification and a back-end computerized system that looks for suspicious spending patterns. Embezzlement is best prevented if several people together are in charge of the books and the books are audited regularly by still more independent people.

The best security systems don't have any single points of failure. When one countermeasure fails, another countermeasure is there to confront the attacker. Relying on a single security countermeasure—a single silver bullet that may protect you absolutely—is a far riskier strategy than relying on a series of countermeasures that can back each other up if one fails—a hail of lead.

In J.K. Rowling's first *Harry Potter* novel, Professor Dumbledore arranges to protect the Philosopher's Stone (the Sorcerer's Stone, in the American edition) not with one but with a series of countermeasures: Fluffy the three-headed dog, devil's snare plants, a locked door with flying keys, a giant chess game puzzle, a troll, a logic problem involving potions, and finally a magic mirror. An attacker would have to penetrate every one of them, rather than just the weakest of them. Security is not just a matter of numerous countermeasures, but countermeasures that work independently and in series and that present different sets of challenges to the attacker. In general, a moat–wall combination is better than two walls.

Defense in depth is important in the real world, too, and ignoring it can have tragic consequences. Near the turn of the last century, because of hot stage lights, theater fires were an extremely serious concern. The Iroquois Theater in Chicago was advertised as "absolutely fireproof" because of a technological innovation: an asbestos curtain that would drop down and shield the audience from a backstage fire. The owners were convinced of the infallibility of their invention and didn't bother with sprinkler systems or other fire safety features. You can already guess how this story ends. On 30 December 1903, an overflow crowd (fire marshals hadn't invented "maximum capacity" rules yet) sees a fire erupt as a backdrop touches a calcium-arc spotlight and ignites. The asbestos curtain drops but gets stuck on another stage light. When the cast and crew flee out the stage door, the wind rushes in and flings fireballs into the audience. The audience is unable to open the doors and flee because some of them are locked; other doors are unlocked, but opening them requires operation of a small lever of a type unfamiliar to most patrons. The death toll was 602.

Certainly this cautionary historical account points out the hubris of the theater owners, who believed that their one security device rendered the need for defense in depth irrelevant. But it also illustrates important security principles we've already discussed:

- The asbestos curtain was designed to drop automatically: This was surely a system of great complexity based on a novel technology, especially considering that this event took place a hundred years ago.
- The system didn't fail in the way the designers expected, but failed because of a surprise interaction with another system: the stage lighting system itself. Clearly the designers were aware of the danger the lighting system posed in igniting a fire. But they failed to anticipate how it could inhibit their countermeasure itself.
- The system failed badly: An initial failure in one system cascaded into failures in other systems. Because of overcrowding, the exit aisles were doubtless jammed. Then the doors through which the audience was supposed to escape didn't open—some because they were locked, some because they employed a new and unfamiliar door-opening mechanism.
- There was a single weakest link; security depended on the asbestos curtain. If it failed, the whole system failed.
- The curtain was nothing more than security theater. Despite all the advertising, the curtain wasn't even made of asbestos; it was constructed of cotton and other combustible materials.

All told, the Iroquois fire was a failure on multiple system levels, but again it illustrates how delusory and dangerous the notion of a single, perfect system really is. Foolproof, infallible, absolutely secure ... sure signs of hucksterism, or worse. Because each security counter-measure is flawed, only defense in depth can provide robust security.

Even in the animal kingdom there are instances of weakest-link security failures due to the lack of defense in depth. By and large, ants differentiate friends from foes by one security system: their sense of smell. Beetles sometimes foil this single-point-of-failure security by sneaking into the ant colony and laying low, playing dead if attacked, until they acquire the scent of their ant neighbors. After that, they're tolerated in the nest by the ants even if they feast on ant larvae.

Defense in depth is standard military defensive doctrine. A mine-field by itself is not nearly as effective as a minefield "covered" by artillery fire. Armored units alone in a city are vulnerable; armor together with dismounted infantry is much more effective. Ground and air forces work best together.

Sometimes defense in depth comes through *overengineering*. Overengineering has to be done carefully because it means more com-plexity, and more complexity often means less security. There are two basic ways to overengineer a system. You can make it more redundant than it needs to be, or you can give it greater capacity than it needs. Airplanes are secured using the first method. There are backup systems for all critical systems, and if the primary system fails or is attacked, the secondary system (or even the tertiary mechanical system) can take over. Buildings are secured using the second method; they are generally designed to take ten times the expected load before collapsing.

The military often engineers much closer to the failure point; if a U.S. submarine hull is engineered to survive depths of 600 feet, then the captain is permitted to dive to 400 feet routinely. In other words, instead of the hull being designed to take ten times the expected load, it is designed to take one and a half times the expected load. The mili-tary makes different security trade-offs than civilians, because it is willing to accept more risks than civilians generally are.

....

Compartmentalization is another way to help secure the weakest link by avoiding single points of attack. The basic idea of compartmental-ized security is to divide assets into smaller pieces and to secure them separately. Doing this limits the damage from a successful attack. It is a kind of divide-to-defend strategy rather than divide and conquer. For example, some travelers put some money in their wallet and the rest in a pouch hidden under their clothing. That way, if their pocket is picked, the thief doesn't get everything. Smart espionage and terrorist organizations divide themselves up into small cells; people know who's in their own cell, but not those in other cells. If one of their number is captured or defects, she can expose only those she knows.

Again, compartmentalization is common sense, and there are many examples available in current events, from good business practices, and from history. Users have individual computer accounts rather than access to the entire mainframe; doors in an office are locked with different keys and not a single key that everyone has; information access in the military is based on clearance plus something called "need to know."

Compartmentalization is the motivation behind a common security system used by urban drug dealers. They don't want to be caught in possession of both drugs and money—possession is a lesser offense than selling, which a large amount of money would suggest. The solution is to divide the responsibility between two people. One stands on the street corner with the money, but no drugs. Someone else, in a less accessible location, has the drugs. The two communicate with a cell phone, and send buyers between them. If the police stop the dealer with the money (the senior member of the team), he has nothing illegal on him. If the dealer with the drugs is caught, well, that's just too bad. (Some gangs have taken to using minors as the drug runners, since they often face much more lenient penalties and have what might be coldly calculated as a less-well-developed sense of self-preservation.)

Manufacturers, too, make use of compartmentalization. Many chemical companies insist that suppliers ship chemicals to their plants with labels identifying the contents only by a code number. Safety data on the shipment is posted with reference to the code numbers only. Most plant employees, and certainly most outsiders, don't need to know what specific chemicals they are handling, only what safety rules they must follow. By compartmentalizing the content information to only those who need to know it, companies make it harder for thieves to target a particular chemical.

Many attacks can be traced to a failure to apply the principle of compartmentalization. Part of the reason for the extraordinary

increase in identity theft is the tendency of a large number of organizations to rely on a single electronic identity. Different companies, for
example, make the problem worse because they all employ the same
identification markers: "mother's maiden name" passwords, home
addresses and phone numbers, and a Social Security number.

Practiced sensibly, compartmentalization can prevent flaws in one
security system from affecting other systems. Compartmentalization
works hand in hand with defense in depth; they both limit the amount
of damage an attacker can do by circumventing or defeating any single
security countermeasure. If bank security is compartmentalized, then a
compromise at one branch doesn't affect the entire bank. In a compartmentalized computer network, a hacker's success against one part
doesn't automatically gain him access to the entire system. Compartmentalization makes security systems more robust, because small failures don't easily become large disasters.

····

Still another technique can complement defense in depth and compartmentalization. A *choke point* is a defensive structure that forces
people, goods, or data into a narrow channel, one that can be secured
more easily. Think of turnstiles at a train station, checkout lanes at a
supermarket, and the doors to your home. Think of the few border
crossings between two unfriendly countries, and the huge line of
people and cars waiting to go through them. As annoying as those lines
are, it's easier to secure a few crossings than it would be to secure the
same border with hundreds of different crossings. Credit card companies use a single processing system to detect fraud, because that single
system has access to more transactions and can more easily see patterns
of misuse. Used correctly, choke points make good security sense.

But choke points work only if there's no way to get around them.
Security guards checking IDs of people entering a building are much
less effective if someone props open the fire door in the back, where
smokers congregate. Firewalls protecting computer networks are less
effective if there are unprotected dial-up modem lines attached to
computers inside the network. (In a surprisingly common error,
employees working from an unprotected home computer dial into a
well-protected company network, instantly making the network as
vulnerable as the home machine.) Requiring all purchasing approvals
to go through one central office fails if one branch office ignores the
procedure. Revolving doors will keep heat in a building in January, but

not if unlocked conventional doors nearby are constantly in use. It's far easier to secure the border between North and South Korea, where there is only one crossing, than it is to secure the long border between the U.S. and Canada, with 113 official crossings (and infinitely more unofficial ones if you happen to own skis, a canoe, or feet).

The problem with choke points, of course, is that by their very nature they tend to get clogged. Moreover, that clog may be either accidental or part of an attack. If all Internet traffic into a company has to go through a single firewall, network speeds can get slow. If every airline passenger has to go through one of three checkpoints at an airport, long lines will form. The reason airports limit gate access to ticketed passengers is not to prevent terrorists from getting to the gates—any attacker can forge a computer-printed boarding pass—but to reduce the number of people going through the choke point so that security personnel can more efficiently screen those who are actually traveling.

Too many systems use choke points instead of defense in depth or compartmentalization. When used together with the other tools, choke points can help ensure that the weakest link doesn't compromise the entire system. Designed properly, choke points have good failure properties. Emergency exits in buildings are normally locked from the outside, forcing everyone through the choke point of the front door. But these exits can be opened easily from the inside in an emergency. This pair of countermeasures combines the best features of a choke point and defense in depth.

.....

In 1863, railroad tycoon Amasa B. Stone, Jr., invented a new form of bridge construction. The prevalent technique at the time was the Howe system, a structure of wooden trusses supported by iron uprights. This system worked because it leveraged the strength of iron without requiring the difficult joinery of an all-wood bridge. Stone decided to build the same style of bridge across the Ashtabula Creek in Ohio, but entirely out of iron and held together by pressure instead of nails (as the wooden bridge had been). This was a fatal flaw, because if one joint went, the whole bridge went. On 29 December 1876, an iron support failed during a train crossing. The entire bridge collapsed under the train, which caught fire, killing ninety-two. (Those interested should look up Julia A. Moore's poem "Ashtabula Disaster," arguably another negative outcome of this whole event chain.)

Most materials engineers use materials and techniques that have been worked with and researched for decades, if not centuries. Often the most important part of an engineer's (and the whole profession's) store of information is the knowledge of what *doesn't* work and in exactly which circumstances. This doesn't mean that engineers are inherently resistant to change. Instead, it means that they have a healthy respect for, and rely on, a huge body of cumulative knowledge and experience: not only how systems work but of how they respond to different conditions, how they affect other systems, and how they fail. Engineers proposing changes to a complex system need a stack of solid and convincing facts on their side, not just a couple of test cases or pilot studies.

It's no different with security. The tried and true is generally much more secure than the new and untested, especially when technologically complex systems are involved. For example, people who design prisons know how to build them securely. They know about cells, walls, fences, and guard towers. Yes, each prison is unique, but the security is basically the same. I don't think anyone would give the time of day to someone who wanted to design a prison around automatic Prizn-Gard™ robots, even if he promised it would be much more secure and a whole lot cheaper.

Unfortunately, complex technological systems often require the new and untested. In the computer world, people believe unsubstantiated claims of security all the time, because systems are so complex and believing is easier than evaluating, especially when the number of systems is continually growing. Still, in these times of complexity, it's best to fall back on what you know.

When you seek medical attention, you don't, as a rule, want a radical doctor who's offering a new and untested treatment. You want the same thing that cured the previous person. Airplane designers know what works and how it works, and don't make random changes in design. (Some aeronautical engineers experiment, but experimental design is a different field.) When designing traffic-control signage, it's critical that it be done to standard, right down to the color spectrum of traffic lights and the font of the letters. If someone designed a totally new kind of parachute, you wouldn't jump at the chance to try it out.

Advances in security happen continually, but slowly. All new ideas should be treated with skepticism initially. Anyone, no matter how unskilled, can design a security system that he himself honestly cannot break, and therefore cannot imagine being broken, and make grandiose claims for it. More important than any security claims are the credentials of the people making those claims.

No single person can comprehensively evaluate the effectiveness of a security countermeasure. There are just too many ways to break technologically complex security systems. Emergent properties of the system affect security in surprising ways. There's no way to prove the security of a countermeasure or a system; it's only possible either to demonstrate insecurity or to fail trying. In mathematics, this is called proving the null hypothesis. Long periods of public evaluation, by motivated evaluators, are the only evidence of security that we have.

If following this principle sounds like it will make you more vulnerable to class breaks, you're right. Systems have gotten so complicated and so technological that most often it is better to stick with the tried and true, risking a class break, than to use something different and potentially much more insecure. On the other hand, it's not always possible or desirable to use old security ideas. The march of technology regularly gives us new security problems, the solution of which requires new ideas. Along these lines, I prefer to see a younger doctor, one who is more likely to know about newer treatments, than one approaching retirement age. But as much as possible, the new ideas build on old and trusted ideas. And every new aspect makes it more likely that the solution will be shown to be insecure.

Simplicity is another engineering principle that supports good security. Complex systems are insecure in part because of their many potential attack points. Simplicity reduces the number and the complexity of attack points. Simplicity reduces the chances that you'll have a security failure, or forget something, or just get it wrong. Attackers know this, too; simpler operations—with the fewest co-conspirators— are more likely to succeed. Complex systems fail in all sorts of different, often unexpected, ways. Systems designers who want to make their system secure should make it as simple as possible. Cut functionality to the bone. Eliminate options. Reduce interconnections with other systems. There is no substitute for simplicity.

But in today's world, simplicity is rarely possible. Society continually demands more options, greater convenience, and new features in products. The economic incentive, then, is for greater complexity. Technological systems are naturally complex. The more technology, the more complexity. Newer systems are, by their nature, less secure than older systems. Often technology requires complexity, but that doesn't mean simplicity shouldn't be a security goal. Albert Einstein supposedly said: "Everything should be made as simple as possible, but no simpler." He could have been talking about security.

••••

Modern systems are also constantly evolving, which affects security. The weakest link doesn't stay the weakest for long. Every change—in how the system works, how it is used, how it interacts with other systems—has the potential to move the weakest link. Make one link more secure, and another one suddenly becomes the weakest. Many improvements bring less additional security than expected, and any assessment of security trade-offs needs to take this into account.

Someone might think: "I am worried about car theft, so I will buy an expensive security device that makes ignitions impossible to hot-wire." That seems like a reasonable thought, but countries such as Russia, where these security devices are commonplace, have seen an increase in carjackings. A carjacking puts the driver at a much greater risk; here the security countermeasure has caused the weakest link to move from the ignition switch to the driver. Total car thefts may have declined, but drivers' safety did, too.

For years, France had a lot of credit card fraud. Telephone lines and calls were expensive, so most transactions were verified by the merchant checking the card against paper booklets of fraudulent account numbers. In 1985, French credit card companies introduced credit cards with embedded computer chips—*smart cards*—as a security countermeasure, and in the first year fraud went down 25 percent. However, fraud in the countries bordering France, especially in the regions of those countries next to France, went up because the criminals simply refocused their efforts. Total fraud in the region remained constant, it just moved around a little. Along similar lines, when your neighbor buys a home alarm system, it increases your risk of burglary ever so slightly.

A month or so after 9/11, I was standing in a seemingly endless queue for airport security. There must have been hundreds of us in a line that snaked back and forth in front of the security checkpoint. The person in front of me remarked how much safer she felt since the airport screeners started doing a more thorough job. I thought: "I don't know. If I wanted to detonate a bomb and kill people, I would do it right here." It was a frightening realization, but a security countermeasure had forced a massing of people *before* security, not after.

Strengthen one link, and another link becomes more attractive to attack. Hence Step 1 of our five-step process—"What assets are you trying to protect?"—and Step 4—"What other risk does the security solution cause?" By limiting the scope of the assets to be protected (airplanes instead of the air transportation system), the government implemented a solution that mitigated one risk but exacerbated another.

Of course, when a security system is in place, the tendency is to use it for as many things as possible. Unfortunately, it's not generally a good idea for a security system designed for one set of circumstances to be used for another. The analysis that led people to believe the system was secure initially may not be valid in the new application. Either there's a new weakest link, or the links are now too weak to defend against the new threats. The problem is that no one does a new analysis, believing the system to be "secure" in some definitive permanent sense.

For example, governments had a security system that protected driver's licenses from fraud. It was based on the threats against a driver's license: who was likely to steal it, what they could do with it, what a driver's license was worth. Then, over the course of several decades, the driver's license came to be used for more things than proving to a police officer that you are permitted to drive: It allows you to write a check at a retail shop, it serves as proof of age at a bar, and it plays a key role in securing permission to board a domestic aircraft. Suddenly all the security trade-offs made about the driver's license— how much security it needs to provide—are no longer valid, because its uses have expanded far beyond its original purpose.

Social Security numbers were originally designed to ease the administration of benefits within the Social Security system. Today they function as national identification numbers in the U.S., without any rethinking of the security surrounding the numbers and the system. Even worse, as mentioned earlier in this chapter, in many applications they function as a kind of secret password. If you know your Social Security number, the system assumes that you must be you.

Other examples of bad security reuse are everywhere. Wooden guardrails on roads might have been good enough when cars chugged along at 30 mph, but they are useless in protecting a driver in a car doing 60. The Internet was originally designed as a medium by which scientists could share information, not for the widespread commercial use we are seeing today. Airports, train stations, and other transportation centers were originally designed to move the maximum number of people through them as efficiently as possible; now they must be redeployed for the antithetical mission of being a front-line defense against terrorists and drug runners. It's not that any of these systems was designed with bad security, it's that they were designed to protect certain assets and are now being used to protect other, more valuable, assets from more dangerous threats.

Many retailers use credit card numbers as account numbers, even though they are not designed for that purpose. In the 1990s, credit card companies wanted to suppress the number in transactions in order to increase security, but retailers complained because they were using the number for customer identification. This kind of conflict is common. As a security system is used over time, people forget why it was created. Security reuse must take into account that the potential attackers change along with the security goals of the system. A password-based security system that might have been good enough to protect personnel files is no longer suitable when it is being used to store million-dollar stock trades. The security it provides is no longer sufficient.

Most security engineering is reactive. People tend to focus on the problem of the moment, and few have the wisdom and foresight to focus on the weakest link. As a result, most of the time new security countermeasures are in response to something that has already happened. Someone I know who sells burglar alarms to homeowners always asks new customers: "Who in your neighborhood got robbed?" Think of all the steps that have been taken *after* 9/11.

Airplane security is a good example. As late as the early 1970s, people were able to take handguns and hunting rifles on planes with them. In 1973, in response to a string of hijackings, metal detectors and X-ray machines screened all passengers and carry-on luggage. In 1985, terrorists hijacked a TWA plane using weapons smuggled past the metal detectors by cleaning staff, who hid guns and grenades in the airplane's bathroom. In response, the FAA mandated that the airlines conduct background checks of ground crews. In 1986, a terrorist hid ten pounds of explosives in the luggage of an unwitting girl-friend—El Al security detected it before it got on the plane—and then everyone was asked whether they packed their own luggage. (The FAA discontinued this practice in August 2002.) Also in the 1980s, terrorists smuggled bombs into their checked luggage and then didn't board the plane. There were two responses: positive bag matching so that a passenger's luggage didn't fly unless the passenger flew with it (Europe decided the trade-off was worth it long before the U.S. did), and X-raying of checked luggage. The 1988 bombers of the Pan Am plane over Lockerbie, Scotland, hid their plastic explosives in a radio in a piece of checked luggage. So now there's a new generation of X-ray machines with technology designed to identify such materials. And after Richard Reid hid a bomb in his shoes, security screeners started asking some people to take off their shoes for extra screening.

New Zealand added metal detection on its domestic flights after 9/11. The country didn't bother with it before because you can't hijack a plane to anywhere: The islands are so remote that planes on domestic routes simply don't have enough fuel.

In 1982, someone added cyanide to Tylenol capsules and put the bottles on shelves in six different Chicago stores; six people died. As a result, packaging specifically designed to be tamper-evident—largely unheard of before—became the norm. Now there is even defense in depth: seals on the outer packaging and on the bottles themselves.

If the goal of security is to protect against yesterday's attacks, we're really good at it. And to some extent, that is one of security's goals. Most attackers are copycats; they're simply not smart enough to respond to changes in systems, functionality, security, and technology and to develop new attack techniques. The Tylenol poisoning incident sparked a wave of copycat tamperings: Lipton Cup-A-Soup in 1986, Excedrin in 1986, Tylenol once again in 1986, Sudafed in 1991, and Goody's Headache Powder in 1992. Several murders were disguised as "random" product tamperings. Reactive security can help us defend against these copycats. But that's not enough; we also need to be proactive to defend against those, like the 9/11 terrorists, who can conceive of and execute new forms of attack.

Example: Security While Traveling

Much of the security advice given to travelers is in the general category of compartmentalization: Wear a money belt and carry half of your money and credit cards there, leave a credit card in your hotel room, make a copy of your passport and keep it somewhere other than with your passport. The idea is that even if someone robs you—either mugs you on the street or burgles your hotel room while you're out—you still have some resources.

When you're traveling, you don't have the same support structures you would when you're at home. You may be in a foreign country. You may not speak the language. All you have with you is your luggage. Robberies are therefore potentially more serious when you're traveling because they're harder to recover from. The proposed solution is compartmentalization: not keeping all your money and credit cards in one place.

Step 1: What assets are you trying to protect? Your money, credit cards, passport, and the like.

Step 2: What are the risks to those assets? Specifically, the risk is robbery. Tourists are more likely to get robbed than locals, largely because they don't know the local area and customs. More generally, the problem is recovering from a robbery. None of the countermeasures listed above makes robbery less likely; they rather make it more likely that a robbery won't be catastrophic. They are countermeasures designed to recover from attacks rather than prevent them in the first place.

Step 3: How well do the security solutions mitigate those risks? Pretty well. Compartmentalization makes it easier for a traveler to recover from an attack, because a thief can't steal everything at once.

Step 4: What other risks does the security solution cause? Minimal ones. There's a greater chance of losing cash or credit cards, because now they're stored in more than one place, for example. But that doesn't seem like a major concern.

Step 5: What trade-offs does the security solution require? Convenience, mostly. Travelers must decide if they are willing to go to the extra effort.

There is no single correct answer in this analysis; it must be left up to the individual. These countermeasures are a good idea, and generally effective, but I usually don't bother because I don't think the risk of theft is great enough. Even in the Third World, travel is much safer than many people think it is. Remember that people get their information from the news, which reports unusual events; they generalize from the few horror stories they've heard. My feeling is that the countermeasures aren't worth the trouble. For many other people, they are.

Traveler's checks are a similar recovery countermeasure and can be analyzed similarly. I don't travel with these, either, but I know many other people who do. These days, credit cards have the same security properties and require no extra effort.

Chapter 9

Brittleness Makes for Bad Security

Because security systems fail so often, the nature of their failure is important. Systems that fail badly are brittle, and systems that fail well are resilient. A resilient system is dynamic; it might be designed to fail only partially, or degrade gracefully; it might adjust to changing circumstances. In general, automated systems—which do one thing very well, but only one thing—are brittle. They don't react well to partial failures, unforeseen attacks, new attack technologies, or attackers who cheat. Similarly, homogeneous systems and systems with an overdependence on secrecy tend to be brittle, and to fail catastrophically if the secret is revealed.

In the first months after 9/11, airport security was slow and methodical. People were scrutinized more carefully. Once in a while, though, something would fail. Perhaps someone would run through the security checkpoint and lose himself in the crowd. Or perhaps a metal detector would malfunction for a few minutes before someone noticed. Then everything would stop. Recently departed planes would return to the gates. No arriving flights would open their doors. Everyone would be forced to leave the terminals. The concourses would be searched for people and weapons, and finally everyone would be let in again, through security, one by one. The resulting delay could paralyze an airport for hours and would propagate through the entire U.S. air-traffic system and beyond.

The problem was that the security system wasn't resilient. It didn't fail well.

Bad security is brittle. I use that word because it describes how something breaks and not how difficult to break it might be. Diamond is the hardest substance known, but it can be cut along its cleavage lines using a simple steel blade and a single hammer blow. The brittleness of diamond is its downfall, so to speak, and its hardness can't save it. Jade is not nearly as hard as diamond, but it's resilient. One of the reasons jade can be carved into amazingly intricate and delicate designs is its resilience; its microscopic structure of interlocking fibers allows it to be manipulated without breaking. Steel-reinforced concrete works similarly, but on a different scale.

All security systems fail sooner or later, but brittle security systems fail badly. Break them at one point, and they break completely. Cause a minor problem, and it turns into a major problem. Cause a major problem, and it turns into a disaster.

A nuclear power plant, for example, is inherently a brittle system. A minor problem—either accident or sabotage—has the potential, however small, to turn into a disaster of epic proportions. Most computer networks are brittle; a single error in the operating system can render an entire computer, even the entire network, insecure. A bunker has lots of concrete and armor plating, but if you can penetrate it you've defeated the bunker. It can't run away. It can't fight back. It's got only one security trick.

Good security systems are resilient. They can withstand failures; a single failure doesn't cause a cascade of other failures. They can withstand attackers, including attackers who cheat. They can withstand new advances in technology. They can fail and then recover from failure.

Think of credit card security. There is always going to be credit card fraud. But the cards are hard to counterfeit, the signature provides some security, the online charge system allows the company to disable cards quickly when they learn a card has been stolen, and the authentication procedures ensure—or come pretty close to ensuring—that only authorized users can activate their cards. What's more, one stolen card doesn't affect the rest of the system. Current rates of fraud in the U.S. are about 6 cents per $100 of transaction volume (down from 7 cents in 1998 and 18 cents in 1992), primarily because of the last two countermeasures. Credit card companies can employ other security countermeasures, but they cost so much and will further reduce fraud so little that the trade-off isn't worth it.

Not all recent gains in security rely on such high-tech methods. On a recent flight, I watched a novel security protocol (to me at least) that addressed the issue of how pilots go to and from the lavatory. After the flight attendant cleared the immediate area and blocked the aisle with a beverage cart, she switched places with the pilot, entering and remaining in the locked cockpit while he used the lavatory. Even though the flight attendant was not, presumably, able to fly the plane, having two people in the cockpit at all times is less brittle than having only one person; it fails better. This is a simple, and effective, security countermeasure. (And every time it happened, the guy in seat 2C wearing the sport coat took off his headphones until the drill was over.)

Another low-tech innovation: My U.S. passport reads, "Make two photocopies of your passport identification page. Leave one copy at home. Carry the other with you in a separate place from your passport." This procedure, long practiced by experienced travelers, is now a recommendation printed on U.S. passports. And it is, like the other examples here, a practice that increases resiliency: When your security system fails and you lose your passport, you can recover more easily.

Resilient systems are naturally more secure, and many of the characteristics we learned about in Chapter 8 make a system more resilient. A system is resilient if it is dynamic, if it can respond to new threats and new attacks. A system is more secure if it is heterogeneous and more resistant to class breaks. What other properties make a security system brittle? Static systems are also brittle. Homogeneous systems are brittle. Systems with too many secrets are brittle. Systems that don't fail securely are brittle. (Examples of all these will follow shortly.) The point here isn't that resilience guarantees security, or that all brittle systems are insecure, only that brittleness contributes to the insecurity of any system. Most security systems that work reliably are resilient, and it's more effort to secure a brittle system because the security countermeasures have to compensate for the system's brittleness. A system's resilience is the single most important security property it has.

In honeybees, normally the queen is the only hive member that can reproduce. This is important—it ensures the survival of her genes—and over the millennia, queens have developed several security systems to protect against attack. For South African honeybees, there are three primary security countermeasures. First, the queen exudes a pheromone that suppresses reproduction in worker bees. Second, worker-laid eggs are eaten by other workers, because they don't have the same scent that queen-laid eggs do. And third, worker-laid eggs contain only one complement of chromosomes and therefore only grow up into males. This is an excellent example of defense in depth. However, its fatal flaw is that it's static.

In 1990, South African beekeepers imported a different subspecies: the Cape honeybee. Normally this wouldn't be a problem; different subspecies generally keep to themselves. But one Cape bee wandered into an African hive and laid her eggs there. And as fate would have it, chance mutations in that one bee made her eggs smell like those of the

South African honeybee. And because Cape honeybee workers produce eggs with two complements of chromosomes and end up being female clones of the mother, the African honeybee defenses failed spectacularly. The Cape eggs were uneaten and grew into adults: female clones with the exact same mutation. After a few generations, her descendants dominated the hive. Today there are billions of clones destroying South African honeybee hives all over the continent. The subspecies just couldn't defend itself against the new threat.

•••••

Let's take a more detailed look at some of the properties that tend to make security systems brittle or resilient.

Automated security is usually static: The system's response to an attack is one thing, and one thing only. It's the card reader at your office door that lets you in only if you have a valid entry card (or if you walk in with someone who does). It's the car alarm that goes off if someone tries to steal the car (or simply jostles it). It's the sprinkler system that sprays water in case of a fire (or if someone hangs something heavy from the sprinkler).

Dynamic defenses that can adapt quickly provide better security than defenses that can perform in only a single way: like the police officer who can react quickly in several ways, and respond to whatever is happening at the time. Being static is the primary vulnerability of land-based nuclear missiles, and why the U.S. has air- and submarine-launched nuclear missiles, as well. Think of the human immune system and how it is able to reconfigure itself to attack a host of invading organisms. (The system could be more dynamic, though; this is the same immune system that causes organ transplant rejection. But as I said, all security systems involve trade-offs.) An organism with a damaged immune system, which makes it more static, will almost certainly be ineffective against a new attacker. Dynamic security can respond to new attacks and attackers, advances in attack technology, and attackers who cheat.

Rifled muskets had a much greater range than the old, smoothbore muskets, and when they were introduced in the 1840s, they rendered short-range Napoleonic frontal attacks obsolete. The superior effectiveness of rifles was first seen during the Crimean War (1853–56), and they proved decisive during the Lombardy–Venetia War—France and Piedmont versus Austria—in 1859. But the lessons had not completely sunk in (or crossed the ocean) by the outbreak of the American Civil War. Napoleonic tactics were still believed to be effective and were employed by Burnside at Fredericksburg, Lee in "Pickett's Charge" at Gettysburg, and Hood at Franklin. All these battles resulted in very heavy casualties. Even the great general Ulysses S. Grant succumbed to the temptation and tried a frontal attack at Cold Harbor, with disastrous results.

In the late 1870s, Grant visited France. A French army officer asked him what he had learned from Napoleon. His reply was that he faced two problems during the war. One was the rifled musket behind earthworks, and the other was moving huge amounts of men and materiel by rail, and that Napoleon had nothing to say on either of them. Static security works best against copycats who repeat the same attacks over and over again; innovation, especially innovation caused by technological advances such as rifles and railroads, requires dynamic security.

One advantage of dynamism is that the asset owner can make security decisions on the fly. The web of an orb spider can be torn apart by the very insects it captures, unless the spider subdues the insects and repairs the web. Some spiders will actually cut large bugs free, trading off the meal in order to reduce the web damage. In a world where attackers can change their tactics, or even their objectives, in mid-attack, dynamic defenses that can react quickly and intelligently are the only answer.

Flexibility in security rules is important, because it leads to more dynamic security. For example, imagine that your neighbor left a copy of her house key with you and told you to let workers in to deliver a new couch. This request seems simple, but complexities abound. If the deliverymen showed up with a couch and a matching chair, you would not hesitate letting the men deliver both. If the deliverymen carried out the old couch and matching chair and put them into their truck, you would similarly not be bothered. If they started carrying out the rest of the furniture, you would call the police. None of this was delineated in the original request, but it is all assumed. That's why your neighbor would not trust a child with this task, even if he could physically unlock the door. That's why a computerized system wouldn't work as well. Neither computers nor children can dynamically react to complex security situations.

We also saw how dynamic security adapts quickly to new attacks during 9/11. The passengers on United Flight 93—the plane that went down in Pennsylvania—were successful in foiling the hijackers because they learned, via cell phone, that the traditional model for hijackings did not apply in their case. They were able to perform an impromptu threat analysis and took it upon themselves to reinvent what it meant to defend against the new type of attack they faced. They themselves didn't survive, but they prevented an attack on a strategic target that would have killed them anyway, plus hundreds or even thousands more on the ground. And they taught everyone else how to counter this new attack.

The concepts of brittleness and resilience have their place in personal financial planning, as well—making investments more secure. Smart investors know the value of diversification. They spread their money among a variety of different investments: different stocks, different bonds, commodities, and certainly a variety of industries and perhaps a variety of countries. The idea is that if something goes horribly wrong (or even moderately wrong) with one of their investments, they won't lose everything. Diversification is a way to reduce the effects of specific investment problems. It's a safety system in financial planning; it makes a financial portfolio more resilient against downturns in the market.

．．．．

Homogeneity is another system property worth thinking about. As a rule, homogeneous systems tend to be brittle because they are more vulnerable to class breaks. Class breaks *are* possible in a diversified system, but they're less likely. Diversity is not always possible, but when you can choose it over homogeneity, it's a smart move. On the other hand, homogeneity is a security risk, but it's often a much lesser risk than straying from the tried and true.

If you look around any modern industrial society, you'll see that most security systems are homogeneous. Virtually every house, every car, and every safe uses one of the same few brands of lock. Every commonly used Internet security product is made by one of a few companies. Every subway turnstile in a city is likely to be exactly the same, and only a few companies make them. And only a very few companies make ATMs, baggage X-ray machines, and just about every other security device you can name. Our society is built on mass production; there's little true diversity in the systems that protect us. (Yet homogeneity is good for interoperability, a classic trade-off.)

The security effect of this industrialized homogenization is that when someone invents a new attack, it's almost always a class break. If someone figures out how to bend his subway fare card in such a way as to get a free ride, he can do it again and again, and he can tell his friends how to do it. Even more serious, he might publish how to do it. (Some people publish vulnerabilities for the notoriety, even if there's no financial incentive.) And every subway station in the city is vulnerable until the problem is addressed.

Most biological security systems are not models of homogeneity, but of diversity and resilience. If a new attack, or a variant on an old attack, appears, then the individuals within a species whose genes are better able to resist it flourish at the expense of individuals whose genes make them more vulnerable. Cape honeybees might be devastating South African honeybees, but it's likely that some South African honeybees will evolve to adapt and survive.

At a higher level of complexity, species can be and are wiped out all the time (in the equivalent of highly successful class breaks). Although diversity alone cannot protect against every kind of attack, genetic diversity is a strong defense mechanism. Smallpox devastated the Native American population after Europeans introduced it onto the continent, but the Native Americans weren't completely wiped out. When cultivation artificially reduces genetic diversity, there is a greater risk of ecological disaster. The Irish Potato Famine of 1845–46 was caused by a fungus (*Phytophthora infestans*). Because only one strain of potato was being planted, and potatoes were being planted on land that couldn't support other crops (and because of a whole host of political issues), the effects of the famine were magnified. Traditionally, Andean potato farmers planted a wide variety of different species—a much more resilient strategy.

While diversity is a highly desirable security goal, it's rarely possible in modern systems. Diversity tends to sacrifice the security of the individual in favor of the security of the population. Think about biological diversity; a disease will kill some individuals, but the species will survive. This isn't much of a consolation if you happen to be one of the individuals killed, and in most human security systems, the individual matters more than the community. This means that security systems designed to secure a population simply don't work in those instances. (It's Step 1 again: What assets are you trying to protect?) When I install a door lock, I want to know that *my* house won't get robbed, not that some percentage of houses won't get robbed and that that will ensure the survival of my neighborhood.

. . . .

Secrecy is another characteristic of brittle systems. Most security systems are crammed full with secrets, or at least things that people believe are secret. They range from the combination of a bank vault to the profiling parameters used by CAPPS (the Computer Assisted Passenger Prescreening System) to determine which airline passengers warrant special screening. The relationship of secrecy to security is subtle and often counterintuitive.

Most security requires secrets of some kind. Secrets that can easily be changed, like a vault combination, are more resilient: They fail well. If the secret gets out and you know it's out, you can easily switch to another secret. You can preemptively (in the absence of any attack) replace the combination and should do so periodically anyway. Passwords and PINs are resilient secrets, as are keys and combinations. (A key is a physical device, but the notches in the key are the secret—anyone who knows the notch positions can make a duplicate key.) The launch codes for nuclear weapons are resilient (and I assume frequently changed) secrets, too.

Secrets that are difficult to change, or global secrets, are brittle; they fail badly. Global secrets are system-wide secrets and can include security procedures, details about vulnerabilities and targets, and access details. A secret door is a global secret. It's secure only as long as the people who know the secret can keep the secret. Once the secret gets out, the door is no longer secure and there's no way to regain secrecy. You can't move the door with any ease.

Some systems are riddled with global secrets. Sometimes these are not even secrets at all, merely misplaced beliefs in secrecy. Airline security, for example, has dozens of supposed secrets: how to get out on the tarmac, how to get into the cockpit, the design of the cockpit door, the procedures for passenger and luggage screening, the exact settings of the trace mass spectrometer that detects the chemical residue associated with bombs, how the autopilot software works, what sorts of background checks are run on maintenance personnel. The security of the airline system can be compromised if any of these secrets are exposed.

One of the consequences of all this secrecy is that airline security is inherently brittle. One group of people knows how the cockpit door reinforcement was designed. Another group has programmed screening criteria into the reservation system software. Other groups designed the various equipment used to screen passengers. And yet another group knows how to get onto the tarmac and take a wrench to the aircraft. All of those secrets must remain secret, and we have to trust everybody who knows any one of them. This doesn't mean that it's impossible to secure an airline system, only that it is more difficult. In general, the more secrets there are and the larger the group of those who share in a secret, the greater the difficulty.

Secrets are hard to keep—stories abound of criminals who are caught solely because they bragged. Secrets are also hard to generate, transfer, and destroy safely. Imagine you're a king, and you want a secret escape tunnel in your castle. It's relatively easy to keep that secret among your trusted advisors, but they're not going to dig the tunnel. In creating the secret, you're left with a whole bunch of workmen who know about it. Of course you could kill the workmen, on the assumption that they hadn't yet divulged the secret. But then what about later, if one of your trusted advisors takes a job with another king? You can't force him to forget the secret. And if you're a CEO and not a king, you can't rely on the chopping block and a large ax to solve your security problems.

A bank safe, on the other hand, has only one (nonglobal) secret: the combination, which the bank's officers set themselves. It doesn't matter if dozens of people know about the safe's existence, if they don't know the combination. It doesn't matter if the bank president takes a job with another bank. The remaining officers can simply change the combination, and the old secret becomes irrelevant.

Remember to change it at least once, though. Safes ship with a default combination. This is an example of a global secret—more a belief in secrecy and less an actual secret. A surprising number of safes in active use still have their default combination, a "secret" also known by quite a few criminals. Even worse, many safes have secret drill points, which only professional safemen, manufacturers, and really good criminals know about. These are spots in the safe that are comparatively less armored than the rest and which, when drilled, allow the lock to be removed or the boltwork to be retracted by manipulating something through the holes.

Hiding a copy of the house key under the mat is another example of a brittle secret. Its effectiveness is predicated on none of the attackers knowing the secret. It's not really a secret at all. It's a belief in secrecy, and a misplaced belief at that.

As much as you should avoid secrecy as a critical component in a security system, it may be your only option if the security system is badly designed. It's little more than a cover-up, so it's always worth asking: "Is the secret necessary for security?" And "How feasible is it that this secret remain secret?" And "What are the consequences when—not if—the secret is revealed?" (The IRS works hard to keep secret the search criteria it uses to flag tax returns for audit.) The fewer secrets a system has, the more secure it is. The fewer people who know those secrets, the more secure a system is. Remember Ben Franklin's adage: "Three may keep a secret, if two of them are dead." The more easily the secrets can be changed by those authorized to change them, the more secure the system.

. . . .

Brittle systems often fail insecurely. Many systems respond to failure by continuing to function, but without the security. Modern credit card systems use online verification systems; the merchant credit card terminal calls back to the credit card company, which automatically verifies the credit card number. This is a good system, but if a store's credit card terminal is down because an attacker cut the line, and the clerk processes your credit card using the old, paper-based, manual system without making a phone call for approval, the end result is a far less secure system.

I call this kind of behavior *default to insecure*: If the secure system doesn't work, the system reverts to a less secure system, rather than maintaining security by shutting down. It's a common security problem in commerce and other low-security systems, because it can be exploited to attack a system. If an attacker wants to use a stolen credit card and is afraid that the merchant will catch him if he verifies the card online, the attacker can cut the merchant's phone line or simply loosen a wire beforehand, forcing the defender to default to the backup manual system.

In another example, a company installs an encryption device to protect data going between two computers in two different offices. Or it installs a secure telephone to protect conversations between two offices. Whenever the encryption is turned on, the local intelligence agency maliciously inserts enough interference into the communications line that it's impossible to use. When the encryption is turned off, it's fine. The attacker—probably one of the governments in the countries involved—is betting that the company will decide it is more important to communicate than to be unable to communicate securely. (Sometimes the company resorts to couriers and face-to-face meetings as a result.)

These sorts of attacks exploit the trade-off that people using the system make. In the first case, the merchant wants your money more than he cares about security. Attacks are rare and profits are more important. He ignores security in order to complete the transaction, because the adverse effects of not completing the transaction are worse than the security risk. (The U.S. Post Office is the one exception to this system that I know of; if the credit card terminal is down, it doesn't take credit cards. But it doesn't have to worry that we'll go across the street to the competition instead.) Depending on the relative risks—Is the store selling greeting cards or fine jewelry?—this might even be the rational security decision. The second example is similar; the adverse effects of not having the conversation are greater than the adverse effects of possibly having it eavesdropped on, whatever the official corporate policy. It's hard to imagine people not having a phone conversation just because the encrypted phone doesn't work. All of us know that cell phones can be eavesdropped on—even digital cell phones—but we use them anyway. Even the military doesn't have the discipline not to communicate if it cannot communicate securely.

We often see the default-to-insecure problem in computerized systems. In computers, and especially in computer networks, backward compatibility is important. That is, a system needs to work with the previous versions of the system that have already been fielded. Think about what would happen if you bought a new version of your word processor and couldn't read any of the files you created with the old version. Sometimes fixing a security problem in a system will make the new system incompatible with the old, insecure, system. If it remained compatible, attackers could exploit the backward compatibility to attack the system.

Smart security designers assume that any security system will fail eventually, and make sure the system will remain secure even in failure. If a slot machine fails, it should not send coins pouring into the payout tray. If a cell phone's security fails, it shouldn't fail by allowing the user to make free long-distance phone calls. If a prison security system fails, it shouldn't automatically open all the cell doors. Or should it? In 2001, twenty-six inmates died in their cells during a prison fire in Chile. These trade-offs can be harder to judge than they appear at first.

"Fail-secure" as we've been discussing it here is akin to the "fail-safe" principle used in safety engineering. If a microprocessor in an automobile fails, we don't want it failing by forcing maximum throttle or making the engine suddenly seize up. If a nuclear missile fails, we don't want it failing by launching. Failing safely and securely are both good design principles. But as we saw in the prison fire example, they can be in conflict with each other.

Example: Increased Secrecy

Since 9/11, there has been a broad trend toward increased secrecy in the U.S., including:

- Information about the inner workings of the military and the Department of Homeland Security
- Potential vulnerabilities in our infrastructure and the countermeasures used to defend them
- Information about people who have been arrested or otherwise detained
- Some scientific and engineering research

The philosophy in all these cases is that the information must remain secret to keep it out of the hands of terrorists. It's a complicated trade-off, and the five-step process sheds some light on the nuances of the issue.

Step 1: What assets are you trying to protect? The threat is terrorism, so potentially, everything is a target. Specific to this example, the asset being protected is information that a terrorist can use to better attack his real targets.

Step 2: What are the risks to those assets? Terrorism. Specifically, the risk is that terrorists will use this information to launch terrorist attacks more easily, or more effectively.

Step 3: How well does the security solution mitigate those risks? The security solution is to keep secret information that terrorists could possibly use. This is an inherently brittle countermeasure and is unlikely to work very well. Many people have to know the secret in order to do their jobs, and the odds are good that one of them will divulge it, either by accident or because he doesn't believe that the secret should remain secret.

Step 4: What other risks does the security solution cause? This is a classic trade-off between short-term security and long-term insecurity. In the computer security world, we've learned that publishing security vulnerabilities is the only way to get software vendors to fix them. Before publication was the norm, vendors would routinely deny the existence of vulnerabilities and not bother fixing them. Systems would stay vulnerable for years. The risk, of course, is that the attackers learn about the vulnerabilities and exploit them.

Another risk is that the users of the security system might not know its true efficacy. This situation occurred recently with regard to door locks. There is a vulnerability in a certain type of door lock that has existed for a hundred years or so. Many lock-
smiths knew about the vulnerability, as did many criminals. But most customers had
no idea about the vulnerability and bought the locks in blissful ignorance.

Even though some lock companies knew about the vulnerability, they chose not
to aggressively market more secure systems. This is a third risk: There won't be a
market for improved security unless people know about vulnerabilities and demand
solutions for them. Even now, those insecure locking systems are still being sold to
buyers who don't understand the vulnerability.

And there's a fourth risk: that criminals will use secrecy to hide their own
actions. An unethical chemical company might keep details of its operations secret
ostensibly in order to keep the terrorists from learning about potential vulnerabilities,
but actually to prevent ordinary citizens or certain parts of the government from learning about environmental violations. It is because of this risk that democratic countries
conduct the business of government in the open.

Step 5: What trade-offs does the security solution require? Most information has
both good and bad uses, and often more good uses than bad ones. When information
is kept secret, we as a society lose both the good and the bad uses of that information. This is a significant trade-off.

It's a complicated trade-off, though. If the power plant down the street were vulnerable to a terrorist attack, you would want to know so that you could make an
informed choice about the neighborhood you live in. If details about exactly how terrorists should attack in order to do maximum damage to the plant and the neighborhood
were made public, you'd probably wish the details had remained unpublished. But if
the details hadn't appeared, you wouldn't have the ammunition to pressure the power
plant to increase security. The trade-off depends on the particulars of the secret.

If a group of people engaged in security research can make use of the secret,
then publishing it is a way to improve security. If there are others who need to build
security systems and who are on your side, it makes sense to publish information so
everyone can learn from the mistakes of others. If there are significant good uses for
the information tangential to security, it often makes sense to publish. If secrecy
adversely affects society in other ways, then it often makes sense not to publish.

Let's take some real examples.

- Imagine a "panic button" in an airplane cockpit. Assume that the system was
  designed so that its publication would not affect security. Should the government
  publish it? The answer depends on whether or not there is a public community of
  professionals who can critique the design of such panic buttons. If there isn't,

then there's no point in publishing. And if all aircraft don't have such buttons, the list of aircraft that have the countermeasure installed should not be published.

- Should the government publish its tactical warfare plans? There isn't a large community of people outside the tactical warfare planning group who can benefit from the information, but there are potential enemies that could benefit from the information. Therefore, it is better for the government to keep the information classified and disclose it only to those who need to know.

- Should the government publish the details of meetings and negotiations? For example, discussions on whether or not to introduce a particular bill or implement a particular policy. Or meetings between government officials and lobbyists. Almost certainly; the added security that comes from government not being allowed to operate in secret is far more important than any minimal security advantage from the secrecy.

- Should the demolitions industry publish recipes for new and more powerful explosives? Should rocket scientists publish detailed engineering information? Almost certainly. This scientific information is vital for scientific research, and keeping this information secret stifles the cycle of innovation and invention that fuels the economic engine of the world's open societies. Sure, keeping it secret would decrease the number of people who could do bad things with that information, but the social trade-offs would be enormous.

- Should the police keep the identities of arrested terrorists secret? Certainly not. The security stemming from such secrecy is negligible—it's reasonable to assume that his cohorts know that he's been arrested, and certainly that he's disappeared. The additional security that comes from forcing the police to release the names of detainees—security from police abuse and police errors—is considerable.

Because the secrecy requirements for security are rarely black and white, each question of publishing has its own trade-off. Does the security benefit of secrecy outweigh the benefits of publication? It might not be easy to make the decision, but the process of making it should be straightforward.

Chapter 10

Security Revolves Around People

Trusted people—people who must be trusted in order for the system to function—are part of any security system. They are a critical element, perhaps *the* critical element, because they're the most resilient part of the system, the most able to improvise, the best equipped to make on-the-spot decisions, and the most skilled at detecting the presence of attackers. But of course human beings, when considered as components of a security system, are a double-edged sword. They can fall asleep, get distracted, and be tricked. They can turn against a security system. A good security system leverages the benefits of trusted people, while building counter-measures to prevent them from abusing that trust.

Security is all about people: not only the people who attack systems, but the people who defend those systems. If we are to have any hope of making security work, we need to understand these people and their motivations. We've already discussed attackers; now we have to discuss defenders.

Good security has people in charge. People are resilient. People can improvise. People can be creative. People can develop on-the-spot solutions. People can detect attackers who cheat, and can attempt to maintain security despite the cheating. People can detect passive failures and attempt to recover. People are the strongest point in a security process. When a security system succeeds in the face of a new or coordinated or devastating attack, it's usually due to the efforts of people.

Here's an example: On 14 December 1999, Ahmed Ressam tried to enter the U.S. by ferryboat from Victoria Island, British Columbia. In the trunk of his car, he had a suitcase bomb. His plan was to drive to Los Angeles International Airport, put his suitcase on a luggage cart in the terminal, set the timer, and then leave. The plan would have worked had someone not been vigilant.

Ressam had to clear customs before boarding the ferry. He had fake ID, in the name of Benni Antoine Noris, and the computer cleared him based on this ID. He was allowed to go through after a routine check of his car's trunk, even though he was wanted by the Canadian police. On the other side of the Strait of Juan de Fuca, at Port Angeles, Washington, Ressam was approached by U.S. customs agent Diana Dean, who asked some routine questions and then decided that he looked suspicious. He was fidgeting, sweaty, and jittery. He avoided eye contact. In Dean's own words, he was acting "hinky." More questioning—there was no one else crossing the border, so two other agents got involved—and more hinky behavior. Ressam's car was eventually searched, and he was finally discovered and captured. It wasn't any one thing that tipped Dean off; it was everything encompassed in the slang term "hinky." But the system worked. The reason there wasn't a bombing at LAX around Christmas in 1999 was because a knowledgeable person was in charge of security and paying attention.

There's a dirty word for what Dean did that chilly afternoon in December, and it's *profiling*. Everyone does it all the time. When you see someone lurking in a dark alley and change your direction to avoid him, you're profiling. When a storeowner sees someone furtively looking around as she fiddles inside her jacket, that storeowner is profiling. People profile based on someone's dress, mannerisms, tone of voice ... and yes, also on their race and ethnicity. When you see someone running toward you on the street with a bloody ax, you don't know for sure that he's a crazed ax murderer. Perhaps he's a butcher who's actually running after the person next to you to give her the change she forgot. But you're going to make a guess one way or another. That guess is an example of profiling.

To profile is to generalize. It's taking characteristics of a population and applying them to an individual. People naturally have an intuition about other people based on different characteristics. Sometimes that intuition is right and sometimes it's wrong, but it's still a person's first reaction. How good this intuition is as a countermeasure depends on two things: how accurate the intuition is and how effective it is when it becomes institutionalized or when the profile characteristics become commonplace.

One of the ways profiling becomes institutionalized is through computerization. Instead of Diana Dean looking someone over, a computer looks the profile over and gives it some sort of rating. Generally profiles with high ratings are further evaluated by people, although sometimes countermeasures kick in based on the computerized profile alone. This is, of course, more brittle. The computer can profile based only on simple, easy-to-assign characteristics: age, race, credit history, job history, et cetera. Computers don't get hinky feelings. Computers also can't adapt the way people can.

Profiling works better if the characteristics profiled are accurate. If erratic driving is a good indication that the driver is intoxicated, then that's a good characteristic for a police officer to use to determine who he's going to pull over. If furtively looking around a store or wearing a coat on a hot day is a good indication that the person is a shoplifter, then those are good characteristics for a store owner to pay attention to. But if wearing baggy trousers isn't a good indication that the person is a shoplifter, then the store owner is going to spend a lot of time paying undue attention to honest people with lousy fashion sense.

In common parlance, the term "profiling" doesn't refer to these characteristics. It refers to profiling based on characteristics like race and ethnicity, and *institutionalized* profiling based on those characteristics alone. During World War II, the U.S. rounded up over 100,000 people of Japanese origin who lived on the West Coast and locked them in camps (prisons, really). That was an example of profiling. Israeli border guards spend a lot more time scrutinizing Arab men than Israeli women; that's another example of profiling. In many U.S. communities, police have been known to stop and question people of color driving around in wealthy white neighborhoods (commonly referred to as "DWB"—Driving While Black). In all of these cases you might possibly be able to argue some security benefit, but the trade-offs are enormous: Honest people who fit the profile can get annoyed, or harassed, or arrested, when they're assumed to be attackers.

For democratic governments, this is a major problem. It's just wrong to segregate people into "more likely to be attackers" and "less likely to be attackers" based on race or ethnicity. It's wrong for the police to pull a car over just because its black occupants are driving in a rich white neighborhood. It's discrimination.

But people make bad security trade-offs when they're scared, which is why we saw Japanese internment camps during World War II, and why there is so much discrimination against Arabs in the U.S. going on today. That doesn't make it right, and it doesn't make it effective security. Writing about the Japanese internment, for example, a 1983 commission reported that the causes of the incarceration were rooted in "race prejudice, war hysteria, and a failure of political leadership." But just because something is wrong doesn't mean that people won't continue to do it.

Ethics aside, institutionalized profiling fails because real attackers are so rare: Active failures will be much more common than passive failures. The great majority of people who fit the profile will be innocent. At the same time, some real attackers are going to deliberately try to sneak past the profile. During World War II, a Japanese American saboteur could try to evade imprisonment by pretending to be Chinese. Similarly, an Arab terrorist could dye his hair blond, practice an American accent, and so on.

Profiling can also blind you to threats outside the profile. If U.S. border guards stop and search everyone who's young, Arab, and male, they're not going to have the time to stop and search all sorts of other people, no matter how hinky they might be acting. On the other hand, if the attackers are of a single race or ethnicity, profiling is more likely to work (although the ethics are still questionable). It makes real security sense for El Al to spend more time investigating young Arab males than it does for them to investigate Israeli families. In Vietnam, American soldiers never knew which local civilians were really combatants; sometimes killing all of them was the security solution they chose.

If a lot of this discussion is abhorrent, as it probably should be, it's the trade-offs in your head talking. It's perfectly reasonable to decide not to implement a countermeasure not because it doesn't work, but because the trade-offs are too great. Locking up every Arab-looking person will reduce the potential for Muslim terrorism, but no reasonable person would suggest it. (It's an example of "winning the battle but losing the war.") In the U.S., there are laws that prohibit police profiling by characteristics like ethnicity, because we believe that such security measures are wrong (and not simply because we believe them to be ineffective).

Still, no matter how much a government makes it illegal, profiling does occur. It occurs at an individual level, at the level of Diana Dean deciding which cars to wave through and which ones to investigate further. She profiled Ressam based on his mannerisms and his answers to her questions. He was Algerian, and she certainly noticed that. However, this was before 9/11, and the reports of the incident clearly indicate that she thought he was a drug smuggler; ethnicity probably wasn't a key profiling factor in this case. In fact, this is one of the most interesting aspects of the story. That intuitive sense that something was amiss worked beautifully, even though everybody made a wrong assumption about *what* was wrong. Human intuition detected a completely unexpected kind of attack. Humans will beat computers at hinkiness-detection for many decades to come.

And done correctly, this intuition-based sort of profiling can be an excellent security countermeasure. Dean needed to have the training and the experience to profile accurately and properly, without stepping over the line and profiling illegally. The trick here is to make sure perceptions of risk match the actual risks. If those responsible for security profile based on superstition and wrong-headed intuition, or by blindly following a computerized profiling system, profiling won't work at all. And even worse, it actually can reduce security by blinding people to the real threats. Institutionalized profiling can ossify a mind, and a person's mind is the most important security countermeasure we have.

···

People are essential for security, but people are a double-edged sword. They are often the weakest security link and the main reason why security fails. People can cause security failures in many ways. Insiders can turn out to be malicious. Attackers can exploit the rarity of failures and play on the tendency of people not to believe in those exceptions. People can also follow the time-honored practice of slacking off, of letting their attention stray from the task at hand to thoughts of ... whatever. The trick is to build security systems that maximize the beneficial aspects of people and minimize their detrimental aspects.

Every security system, without exception, needs *trusted people* to function, though these people are not necessarily trustworthy. The person who installs your front-door lock is, by definition, a trusted person. You have no choice but to trust him, but he could make a copy of your key for himself, and you wouldn't be the wiser. In a hospital security system designed to protect patient privacy, doctors are trusted. They have to know a patient's private medical background. Other trusted people are guards watching surveillance cameras, people writing parking tickets, airline baggage screeners, customs inspectors like the three in Port Angeles, and police officers who respond to cries for help. We don't have to like it, but we have no choice but to trust these people. Without them, systems would not function.

Ron Harris was a computer lab technician for the Nevada Gaming Control Board. His status as a trusted insider meant that he could rig slot machines undetectably. Between 1992 and 1995, he modified the software in some of these machines so they would pay jackpots when a certain sequence of coins was played—say, three coins at the first pull, at the next pull one coin, then two coins, then five coins, and so on. The same sort of thing happened in 2002; Chris Harn was a senior programmer at Autotote, which processed computerized horse-racing bets. He used his insider status to win $3 million betting on the Breeders' Cup. The only reason both of these people got caught is because they were overly greedy or stupid; the bets they placed were so large that they aroused suspicion. A smarter attacker would have gotten away with these crimes.

The more trusted people a system employs, and the more we must trust them, the more brittle it is. Because people are the weakest link, the fewer who must be trusted in a system, the better. The more complex a system is, the more trusted people it typically has. For example, let's consider a centralized database of tax records that officials can access when auditing returns. This system clearly contains enough sensitive information to make security important. Think about the number of trusted people who might be required to set up and run it.

Anyone who must look up people's personal tax records to perform her job must be trusted not to abuse the system or to disclose what she sees. In the U.S., IRS employees are routinely discovered and punished for looking up the tax records of their friends or of celebrities, sometimes for money. Employees have also sold documentation about tax audit procedures to help unscrupulous people reduce their risks of being audited. (In similar abuses, police officers have been known to sell computerized criminal records to criminals, and private records about sick people have been obtained from health services organizations and sold to drug companies.) All the people who enter data into the database must be trusted. They must be trusted to enter only valid information and not to disclose any information they see. (If there's a system to catch errors and correct them, anyone involved in that system must be trusted, as well.) Everyone who programs the database must be trusted, both to build a system that functions correctly and to build a system that has no back doors or security flaws— intentional or accidental. There have been cases where a programmer entrusted with building the corporate payroll system included a feature that would periodically give him a raise, automatically. Everyone who maintains the system—the computers, the networks, and the database—must be trusted. In addition, if paper copies of records are printed out, the people who have access to the file cabinets and take the trash out must be trusted. If the paper trash is put in locked bins to be shredded, those doing the shredding must be trusted. And even the people who eventually dismantle the system have to be trusted. That's a lot of trust, and what would be required for a tax database would be required for any similar database system.

As I've noted, trusted people, because of their insider access, can perform attacks that outside criminals might find difficult or impossible. The CEO and the CFO of a company can collude to cook the books, because those are the very people entrusted with making sure that the books are proper. (This is the point of an outside auditing firm, which works only if the outside auditor truly is independent; witness the Enron accounting scandal, which also encompassed the Arthur Andersen accounting firm.) Eight members of the 1919 Chicago White Sox could conspire to throw the World Series to the Cincinnati Reds, because they were the very people trusted to play fairly.

Sometimes machines—trusted systems—are used in lieu of people. This option doesn't reduce trust, it only shifts it. Consider a mechanical voting booth with levers. Instead of trusting actual vote takers, in effect we have to trust all the people who designed, built, installed, and verified the machine. The mechanical system is more secure, because there are fewer people to trust. But at the same time it's much more dangerous, because a single untrustworthy person can do much more damage; a class break is easier. If the designer (or repairer, or installer) of the voting machine wishes to and has the right skills, he might tinker with internal gears and levers to have the machine record two votes for every one vote actually cast for his party. And if the voting machine is electronic—that is, if it runs on software—his manipulation will be all the easier to accomplish, and all the more difficult to detect or trace.

. . . .

Trusted people are everywhere. In a company, every employee is a trusted person who is allowed inside the company offices, able to use company resources and authorized to act, however broadly or narrowly, in the company's name. In a store, customers are trusted people by virtue of the fact that they are allowed inside the store and can handle the merchandise. In a transit system, every rider is a trusted person who is allowed through the barriers and onto the trains and buses. And all security systems need trusted people to function: Often they need trusted people for their operation, but even automatic systems have trusted people—those who set them up, maintain them, and fix them when they break.

There are three basic ways to secure trusted people—or trusted machines, for that matter. The first: Try to put trustworthy people in positions of trust, and try extra hard for positions of extreme trust. This is why background checks or drug testing are done on key employees. This is why certain jobs are not open to people with criminal records or to noncitizens. This is one of the reasons some jobs require training, professional certification, or membership in a professional society. We can argue about the efficacy of these countermeasures—for example, I can understand why the FBI doesn't hire non-Americans as agents, but why can't non-Americans be trusted with airport security?—but however we as individuals might disagree on specific rules, the general idea is to certify, background-check, and test individuals in order to increase the chance that a trusted person is also trustworthy.

Luckily, most people *are* trustworthy. As a society, we largely rely on people's basic beliefs, ethics, and feelings about what's right. There are also rewards and sanctions, including reputational consequences and a legal infrastructure, that help keep people in line who otherwise might stray.

Unfortunately, not everyone is trustworthy. And even the most trustworthy people can be tempted by opportunity, overcome by emotion, or coerced by others. Someone who won't betray her company's trust for $500 might do it for what she thinks is a safe way to steal $500,000.

The second way to secure trusted people is through compartmentalization. Trust is not black and white, meaning that someone is either trusted or not. And trust is not monolithic, meaning that trusted people are not all trusted to the same degree and in the same contexts. We can increase security by limiting our trust: Give trusted people only the information, access, and compatibilities they need to accomplish their tasks. In the military, people are told only things that they "need to know," even if they have the requisite clearance. Not all company employees are trusted with the same capabilities, information, and access. Transit riders are not trusted to the same degree as transit employees, and station attendants are not trusted to the same degree as transit police officers. Everyone in the world may be trusted to some degree, but not all are trusted—nor should they be—to the same degree. In fact, most countermeasures are put in place to protect a system from insiders trying to exploit their trusted status.

And the countermeasure is not simply a matter of people higher in the hierarchy always being granted a higher degree of trust. Only authorized armored-car delivery people can unlock ATMs and put money inside; even the bank president cannot do so. Station attendants are not trusted to drive trains, and drivers are not trusted to staff ticket booths. You might have to trust a nanny with the health and safety of your children, but you don't have to give her your credit card. We can give some of the people involved with the tax-record database the ability to read information, but not to modify it. Think of a trusted person as having a sphere of trust—a set of assets and functionality he is trusted with. Make that sphere as small as possible.

Limiting trust is central to the debate about arming airplane pilots. Just because a person is entrusted to fly a plane doesn't mean he should automatically be trusted with armed defense of that plane against a hijacker. Dividing the trust—adding a sky marshal with a gun and a lot of close-quarters firearms training—is more secure and more resilient, as it is a form of defense in depth. But that solution requires more people and is more costly. That's the trade-off.

Trust should not only be limited, it also should be granted directly. If you trust Alice, and Alice trusts Bob, that does not automatically mean that you trust Bob. You might not even know Bob. In systems where you do automatically trust Bob, trust is said to be *transitive*. And when trust must be transitive, it creates brittleness. Once some-one enters the airport system, for instance, every other airport auto-matically trusts him. If he's flying from Des Moines to Chicago to San Francisco, he goes through security only once ... at Des Moines. (Some airports rescreen international connecting passengers.) If he gets a weapon into the system, it can be moved around and remain in the system indefinitely. He could spend months slowly sneaking weapons into the airport system and attack the aircraft only when he has a sufficient supply. We know that some of the 9/11 terrorists entered the airline network in Portland, Maine, presumably because they thought security was more lax there. We have no idea if they had a cache of weapons stored in an airport locker, a cache slowly brought into the airport system over the previous months.

Sometimes trust is transitive, but limited. If I'm having a party in my home, I often allow a friend to bring a friend. This works because I know the limitations of the transitive trust; I don't trust this friend's friend to the same degree that I trust my friend, but I trust him to behave himself at the party and not walk off with the flatware.

The third way to secure systems with trusted people is to apply the principle of defense in depth: Give trusted people overlapping spheres of trust, so they effectively watch each other. Imagine how a system to catch and fix errors in a database might work. To make a change in the database, two people have to sign the form. One person makes the change, and then another person checks to make sure the change has been made accurately. A third person periodically reviews selected database accesses to make sure there's no abuse. This is what independent auditing firms are supposed to do for corporations; corporate officers generate the accounting figures, and auditors make sure they're accurate. Checks and balances, sometimes called separation of duties.

In 1995, Barings, the UK's oldest merchant bank, suffered a spectacular attack by ignoring this principle. Nick Leeson was the bank's star Asian currency trader. When currency fluctuations after the Kobe, Japan, earthquake left his funds short of cash at a crucial moment, he disguised his losses by moving them to different accounts. Because he had sole control of both the trading desk and the back office, no one noticed his actions until he had run up losses of £800 million, almost the entire assets of the bank. Barings collapsed, and was bought by a Dutch company for £1.

In 1985, John Walker, a retired U.S. Navy warrant officer, was convicted of selling government secrets to the Soviet Union. He was a security officer, the person responsible for making sure those secrets stayed secret. One of the operational changes the Navy made after his conviction was to expand the practice of two-person control for guarding many of their secrets. Documents are now stored in a safe with two combination locks. One person knows one combination, and another person knows the other. To open the safe, both people have to be present. This practice recalls a technique that was used when the Koh-i-Noor diamond was being transported from Bombay to England in 1850. The diamond was placed in an iron chest, which was double locked and then placed in a larger double-locked chest, with each key being held by different people. (Looking at it another way, all this did was reframe the problem from stealing a diamond to one of stealing a large iron chest and opening it later. But the iron chest was much heavier.)

The vault at Fort Knox works in a similar fashion; no one person knows the combination, but several people know pieces of the combination. It is less likely for one person and his assigned partner to betray their country or steal the gold than it is for one person who can act alone. Similar security countermeasures are in place in nuclear missile silos: To launch a missile, two people have to turn keys and unlock the system simultaneously. And the keyholes are far enough apart that a single rogue soldier can't turn both keys alone. It's just like a corporate checking account that requires two signatures on high-value checks: No one person can break the security.

"Less likely" does not equal impossible, of course. One of the people Walker recruited was his partner; thus he was able to defeat the two-person control. (The old German Democratic Republic solved this problem by ensuring that border guards never worked two shifts with the same partner.)

Many ticket-taking systems employ the same kind of countermeasures. You buy a movie ticket from one person and then walk twenty feet to another person, who takes your ticket, tears it in half, and lets you into the theater. This isn't a full-employment program; it's to limit the amount of trust placed in each person. If there were just one person taking the money and allowing admission, she could easily cheat. She could pocket your money, not give you a ticket, and let you in. (The tickets are there so management can determine how many people have passed through, and then check the contents of the till and ensure that no money has been stolen.) Cheating is much harder if the money collector and the ticket taker are separate people; they have to be accomplices to steal money undetectably. Since you can buy tickets from only one booth, fewer people need to be trusted with the money. (And the money can be secured more easily, because its centralized storage is a choke point.) A dishonest ticket taker could allow friends in for free, but that's a smaller risk to take because no actual money is lost—an example of limiting trust.

Not only can trusted people fail to be trustworthy, and attack the underlying system, but they can inadvertently harm security. Many attacks specifically target trusted people. People can make mistakes; they may not be paying attention. They can fall into bad habits that attackers can exploit. Trusted people can even inadvertently help attackers. They are particularly vulnerable to attackers pretending to be other trusted people, either on the phone or in person and wearing the correct uniforms. For example, building guards can be fooled by an attacker wearing an appropriate uniform; guards have even been known to help properly dressed criminals carry stolen goods out of the building. Computer users can be fooled by someone on the telephone claiming to be from their company's tech support office. In the computer world, this is known as *social engineering*, getting an unwitting insider to help with the attack. In some ways, this vulnerability is a result of the corporate culture of helpfulness and openness; it's a trade-off organizations make.

In 1994, a Frenchman named Anthony Zboralski called the FBI office in Washington, DC, pretending to be an FBI representative working at the U.S. embassy in Paris. He persuaded the person at the other end of the phone to explain how to connect to the FBI's phone-conferencing system. Then he ran up a $250,000 phone bill over the course of seven months. This is nothing more than traditional fraud, but technology makes the attacks easier. Zboralski did not have to risk visiting an FBI office himself; he simply used the telephone. If the person he spoke to became suspicious, he could simply hang up and try again another day.

Convicted hacker Kevin Mitnick testified before Congress in 2000 about social engineering. "I was so successful in that line of attack that I rarely had to resort to a technical attack," he said. "Companies can spend millions of dollars toward technological protections, and that's wasted if somebody can basically call someone on the telephone and either convince them to do something on the computer that lowers the computer's defenses or reveals the information they were seeking."

Naïve computer users are being persuaded by official-looking e-mails to give passwords to their accounts to criminals. Job hunters are being lured by fake job listings into providing all the information someone would need to steal their identities. Some years ago, *Spy Magazine* sent famous people $1 checks, just to see if they would deposit them. Many did, providing copies of their signatures in the endorsements. An attacker could get a copy of someone's signature simply by mailing a check or sending a package they need to sign for and then requesting a copy of the signature form from FedEx. There's no real way to defend against these attacks; they're an effect of a single signature being used to regulate access to a variety of systems.

Social engineering will probably always work, because so many people are by nature helpful and so many corporate employees are naturally cheerful and accommodating. Attacks are rare, and most people asking for information or help are legitimate. By appealing to the victim's natural tendencies, the attacker will usually be able to cozen what she wants.

. . . .

People are critical in any security system. They're the trusted people delegated to enforce the security policy and the trusted people delegated to manage trusted systems designed to do the same thing. The basic idea is to give people the ability, authority, and tools to do their jobs, while at the same time auditing them regularly to ensure that they don't abuse their power.

This sort of trust doesn't lend itself to minimum-wage security personnel, but that's far too often what we get. Unfortunately, almost no one is willing to offer these people pay that's commensurate with their true value or to give them the autonomy they need to do their jobs properly. But, like everything else, there's a trade-off. Good people are expensive, and cheap people are unreliable.

Turning people into drones might work for some low-level security applications, but it won't work against smart attackers. Vigilant human intervention is critical for effective security. Automatic security is necessarily flawed; only people can be dynamic and resourceful and can recognize and respond to new threats and attacks.

The more modern approach is to replace people with technology, and to believe that technology can somehow solve the problems of security. Whether it's national ID cards, or biometric scanners in airports, or computers that automatically search every e-mail for signs of illicit activity, many people believe that all it takes is the suitable piece of technology implemented in just the right way and... abracadabra ... everyone is secure. There is, of course, something intriguing, beguiling about the idea. One of the consistent promises of technology is that it will replace human drudgery. This idea has extra appeal in security, because if it works, all ambiguity and tendency toward failure embodied in human nature could be excised.

But when we replace people with technology, we lose their creativity, their ingenuity, and their adaptability. When we constrain people so much that they become drones, we create the same effect. Technological solutions have an unrealistic tendency to idealize people and their actions. They often assume that people will act as perfect drones, always doing what they're supposed to do and never doing what they're not supposed to do. In reality, people aren't nearly that reliable, but at the same time they're not nearly that rigid. Technological solutions are brittle, and they don't fail well.

Good security uses technology, but centers around people. Good security systems are designed to maximize the value people can offer while at the same time minimizing the vulnerabilities inherent in using them. People are dynamic, and better able to react to new threats and respond to new situations than technology is. They may make certain kinds of mistakes more often than machines, but only people can change the way they react as an attack happens. This realization isn't anything new. The Great Wall of China didn't impress Genghis Khan.

"The strength of a wall depends on the courage of those who defend it," he supposedly said. A wall is static; people make it dynamic.

A vigilant guard watching over a public square is more resilient than an automatic defense system. Computers may be better at details, but human intelligence is less brittle. Guards can react to something they've never seen before: a new attack, a new threat, a new vulnerability. Auditors with a trained eye poring over a company's books provide more resilient security than a computer program does. In any security situation, a person is ultimately doing the attacking. If there isn't also a person doing the defending, the attacker has an enormous advantage.

Computers, however, excel at dull, repetitive tasks. They never get tired, and they are consistent in their responses. It might be easier to fool cameras than guards, but it's harder to catch cameras on a bathroom break, bribe them, drug them, or kick them in the shins.

Security is not about technology. It's about risks, and different ways to manage those risks. It's not a product; it's a process. Technology is an assistant—often a valued one—because it's static and brittle. It can respond only to the predictable. People are dynamic and resilient, and they fail well. The military knows this. With all of its high-technology weaponry, it still relies on people to make the decisions. Even as late as the Iraq War in 2003, soldiers on the ground directed precision bombings.

Therefore, when you evaluate a security system, examine how the people involved are utilized. Are the people involved compensated well? Are they trained well? Do they seem to have a passion for what they're doing, or is it just a job to them? There's never any guarantee, but a well-compensated, well-trained, empowered, engaged, and interested person is less likely to fail. The Transportation Security Administration (TSA) does a much better job at U.S. airport security than the old airline-paid contractors they replaced, primarily because they're better paid, better trained, and better treated. They're professionals, and they generally act as such.

Security technologies are important, but only inasmuch as they support people. Good security systems usually involve technology and people working together, but the people have to run the technology, not vice versa. To make security work, we have to make sure the people understand the technology, what it does and doesn't do, how it works, and how it fails.

Chapter 11

Detection Works Where Prevention Fails

The ideal of any security system is to prevent an attack, but prevention is the hardest aspect of security to implement, and often the most expensive. To be practical as well as effective, almost all modern security systems combine prevention with detection and response, forming a triad that operates as an integrated system to provide dynamic security, resilient failure, and defense in depth. Audits (retrospective detection) and prediction (prospective attempts at detection) don't produce the decisive real-time results that most people associate with security systems, but are extremely important in evaluating and thinking about ways to improve security systems.

Built in 1936, the U.S. Bullion Depository at Fort Knox consists of 16,000 cubic feet of granite, 4,200 cubic yards of concrete, 750 tons of reinforcing steel, and 670 tons of structural steel. The outer wall of the depository is constructed of granite and lined with concrete. Inside is a two-level steel-and-concrete vault with a 20-ton door. The vault casing is constructed of steel plates, steel I-beams, and steel cylinders laced with hoop bands and encased in concrete. The vault's roof is of similar construction and is independent of the depository's roof.

A barrier is the simplest countermeasure. It's a wall, a box, an envelope, or a mountain range. It's something that prevents the attacker from getting to the assets behind it, and it can be an excellent defense. Soldiers in concrete bunkers are harder to kill. Gold stored in Fort Knox is harder to steal. Moats made it harder to attack castles. Skin protects the body. A turtle's shell makes it harder for a predator to bite. The Three Little Pigs didn't sally out to tangle with the wolf; they stayed within their houses. (They did, however, learn valuable lessons about the relative effectiveness of straw, sticks, and bricks as preventive barriers.) Satellites are protected by another type of barrier: miles and miles of difficult-to-travel-in, impossible-to-breathe-in space. Prevention is the first and foremost role of any barrier.

Prevention is passive; it works all the time without anyone having to do anything. Unlike guards, who have to remain alert during their shifts and are relieved afterward, walls are simply there. And not all preventive countermeasures are physical barriers. Photocopy machines are built to include all sorts of technologies to prevent various currencies from being copied properly. Computer software is sold with embedded countermeasures, in the form of hidden codes, to prevent the software from being copied.

Prevention is what everyone thinks about first, but it's actually the hardest aspect of security to implement successfully, and often the most expensive. Suppose you run a company and you want to prevent your key executives from getting kidnapped when they travel to dangerous countries. You can house them in fortress-like hotels. You can hire armored cars for them to travel around in. You can hire bodyguards, working in round-the-clock shifts. You can do all of these things and more, but if a guerrilla army wants to kidnap your key executives, they'll still be able to do so. You can make it harder for the kidnappers, but you can't stop them totally. You simply can't buy enough prevention, and the trade-offs are too great.

The U.S. government spends an enormous amount of money protecting the President from assassination. It's not wasted money, but clearly it doesn't prevent assassinations; witness John W. Hinckley, Jr., a lone operative with no experience whose failure to kill President Reagan in 1981 was a result of his poor aim and not any security countermeasure. Egyptian president Anwar Sadat was assassinated that same year. Since 1995, the leaders of Sweden, Israel, Niger, Congo, and Serbia have been assassinated. Don't think that those countries had lackadaisical security forces protecting their leaders. It's simply that absolute protection, twenty-four hours a day and seven days a week, is impossible.

Prevention is best combined with detection and response. A safe provides defense, but it works best in concert with alarms and guards. The safe—including its walls and the lock on its door—is a preventive countermeasure. In contrast, the alarms are detection countermeasures, and the guards are responsive countermeasures. You cannot make a good decision on how strong a safe you need until you know what detection and response mechanisms will be in place. If guards patrol the offices every hour, then the safe only has to withstand attack for a maximum of sixty minutes. If the safe is in an obscure office that is only staffed during the day, then it has to withstand attack for sixteen hours: from 5 P.M. until 9 A.M. the next day (much longer if the office is closed during holiday weekends). If the safe has a motion sensor attached to it, and the guards come running as soon as the safe is jostled, then it only has to survive attack until the guards arrive—perhaps only minutes.

And safes are sold this way. A safe rated as TL 30 can resist a professional safecracker, with tools, for at least 30 minutes. Another, rated TL-TR 60, means it can resist the same safecracker, with tools and an oxyacetylene torch, for 60 minutes. These ratings are for a sustained attack, meaning that the clock is running only when the safe is actually under attack by professionals with access to the safe's engineering drawings. No secrets, except for the combination. What the safe does is buy time: 15 minutes, 30 minutes, 24 hours. This time allows the various alarms to sound (detection) and the guards to come running (response). Without detection and response, it actually doesn't matter whether your safe is rated TL 30 or TL-TR 60; the burglar will eventually break in and steal whatever is in your safe. An old master locksmith once told me: "Our job is to slow 'em down or make 'em make a lot of noise"—that is, buy the time needed for detection and response to take over.

Fort Knox is no different. Even with the impressive preventive barriers, the entrances to the Fort Knox Depository are under constant guard. Only authorized personnel are allowed inside. There are no tours of the Depository, and visitors are prohibited. (One group of visitors was allowed in once, in 1974, mostly to prove that the site was not a hoax—the only exception I know of.)

Far too often people think about security only in terms of prevention; in fact, detection and response are not only more effective, but also often cheaper, than additional prevention. It's cheaper to add an alarm system to a home than it is to reinforce doors, windows, and walls. And detection can make up for shoddy prevention. If the police can respond to an alarm within seconds, you don't need a very good door lock. If you have enough guards, you can take your money and safely dump it in a pile on the street. If rabbits are alert and numerous enough, they can survive by running away even though they don't have armor or scales.

Banks do not believe that because they have a vault, they don't need an alarm system. Museums don't fire their night guards because they have massive doors and extra-sturdy window locks. Walls, locked doors, and barbed-wire fences are all good protective security measures. It's just that they're brittle, and you need to think about what to do after they fail. Prevention, detection, and response systems are a triad; in concert, they provide dynamic security, resilient failure, and defense in depth. Weak prevention means that you need better detection and response. And without good prevention, detection countermeasures can be more easily subverted. An ounce of prevention might be worth a pound of cure, but only if you're absolutely sure beforehand where that ounce of prevention should be applied. Otherwise, detection and response, taken together, are much more valuable than prevention.

．．．．

Detection can be basic: a guard walking around a warehouse at night, looking for prowlers. Sometimes it's much more complicated: systems of alarm sensors and alerting mechanisms. At the Nuremberg trials, war criminals were locked in cells, and guards watched them twenty-four hours a day. Although this was more to prevent suicide than escape, and it still wasn't 100 percent effective—Hermann Goering managed to kill himself anyway, chewing on a smuggled vial of cyanide—the guards were a good security countermeasure.

Sheep are slow and tasty, and therefore must remain constantly alert. They lie down and rest often, but they sleep in short bursts totaling only eight minutes of deep sleep a day. And their eyes are on the sides of their head, giving them about 340 degrees of vision. All this is to enable them to detect attackers and run away before it's too late.

Detection countermeasures typically are attached to the system being defended—an alarm is the most obvious example of this. But it is possible for detection to be attached to the attacker, instead. In Aesop's fable "Belling the Cat," a group of mice consider a plan to place a small bell around a cat's neck. Think of this as an alarm permanently attached to the attacker; it will warn defenders of the attacker's approach. Implementation was the problem for the mice, however; they needed a suicide beller. (Actually, the problem with belling a cat is that it mitigates the wrong risk. Cats tend to lie in wait and pounce, so a bell is almost totally ineffective. By the time it rings, it's too late.)

We're starting to see effective attached detection systems in modern security. Intelligence organizations have been implanting tracking devices on cars for a long time, and—as the price came down—police forces have been doing so, as well. LoJacks are tracking devices used to find stolen cars. Criminals have been tracked using the OnStar navigation system. Cell phone tracking is sophisticated enough for this kind of surveillance, and at least one company has marketed a tracking device that parents can attach to children. It's easy to imagine enforcing a restraining order by attaching a small tracking device to the restrained person and sounding an alarm whenever he gets too close to a device held by the person he's not allowed to approach. Some people, of course, are able to figure out how to remove tracking devices, or they leave their cell phones at home. Criminals have been known to add a switch to their car that allows them to turn off their taillights, making it harder for the police to track them at night.

Sometimes detection and response systems are deliberately made obvious: A guard wears a uniform and stands where everyone can see him, his weapon obviously visible; a home alarm system comes with a warning plaque you're supposed to display prominently on your lawn. This is detection also serving as prevention, but there's a trade-off. On one hand, a secret detection system is more effective because attackers don't know where it is. An anonymous sky marshal on an airplane, rather than a uniformed one, would be more effective against would-be hijackers. Unidentified in-home alarm sensors—pressure plates, electric eyes, motion sensors—can surprise a burglar who doesn't know they are installed. On the other hand, a hijacker who sees a sky marshal on a flight might decide not to hijack the plane, and a burglar who sees that alarm sensors are employed might decide not to rob the house.

Minefields are another detection system that are often made visible intentionally. To be effective, a minefield doesn't have to be that dense, as long as it makes its existence known. This is why armies post "Danger: Minefield" signs. This, of course, assumes attackers can read and that no one has stolen the signs. Visible detection systems are more effective against a random attack than a targeted attack. A clearly announced home alarm system is meant to keep the owner's home safe; the owner is less concerned about a neighbor's house. (He might not want to advertise the exact placement of some of the sensors, but he certainly wants to advertise the general idea.)

That is not the case with sky marshals. Their goal is to prevent hijackers from succeeding on any plane. Of course they want to stop a hijacking on the plane they are on. But also high on their list is the idea of not giving hijackers another chance. Being incognito gives the marshals a better chance of taking hijackers out of circulation permanently. The preventive value of an identifiable sky marshal is minimal, and probably inconsequential, when compared to the benefit of making the response capabilities more effective.

El Al manages to get the best of both worlds. It publicly advertises that there are two sky marshals on every flight. (With only 90 flights a day, this doesn't represent a tremendous expense.) At the same time, the airline keeps their identities secret from the other passengers on the plane. And since this fact is such common knowledge, El Al doesn't actually have to have two sky marshals on every flight. This is an example of a *random* detection/response system, one that takes place occasionally and is not constant.

On the Berlin subway system (and many others in Europe), tickets or tokens or fare cards are not required to board a train. It's not like London, Paris, New York, or Tokyo, where the turnstile needs to authenticate your fare payment before allowing you through. Berlin passengers can choose to ride without paying, even though payment is the law. Once in a while, transit policemen go through a train and makes sure everybody has a ticket, and anyone caught without a ticket is fined. That's random detection.

For random detection to work, the punishment must exceed the crime or, at least, be perceived as exceeding the crime. If a ticket costs $1, and the police check a passenger once every hundred rides (on the average), a rational and unethical passenger will cheat and pay the fine if caught, but only if the fine is less than $100. It's simple economics. This is not to say that lax punishment motivates people to cheat—most people are ethical—but lax punishment will not deter potential cheaters from cheating. Serious fines, a day in prison, or public humiliation (being on a publicized list of cheats) might serve as deterrents.

International customs inspections fail in this very way. It's impossible to inspect every truck, every shipping container, and every airplane passenger entering a country. Customs officers spot-check and fine people for cheating. But if the fines are both rare and affordable, considering the odds of being stopped, this is not effective random detection, and criminals cheat all the time. Think of the economics of drug smuggling from a "manager's" perspective; it's okay if some smugglers are caught, as long as enough others get through so that the business remains profitable.

For randomized detection to be effective, it must be impossible to avoid the response once you've been selected. Once a friend of mine was traveling by subway in Amsterdam without a ticket. At his station, at the top of a long escalator, he noticed two uniformed policemen checking people's tickets. So instead of taking the escalator, he found another exit from the train station that was unguarded. This was hardly optimum randomized detection. It would have been far better for the entire station to be checked, and for passengers not to know about it until it was too late to find another exit or board another train. The guards, for example, should have posted themselves somewhere that the passengers could not see them until they had already gotten on the escalator, and it was too late to turn back. In many European cities where there is random detection, tickets are often checked while the trains are moving, which is another good method.

This problem with random detection exists in many U.S. airports. It's easy to bypass the "special screening" procedures by only going through security checkpoints when they're busy. This won't work if your boarding pass is marked for special screening (although with Internet-based check-in you can look at your boarding pass and find out if you've been selected as much as twenty-four hours before your flight), but it will work against the random checks. Since I started paying attention to the habits and cadence of the security screeners, I've never been singled out at an airport for "additional screening."

  ••••

Failures are a critical problem for detection systems. Passive failures are obvious—when the system fails to detect an attack—but active failures, when the system signals a false positive or false alarm, can be even more significant. Some false positives are commonplace and merely irritating, as when car alarms go off even when there is no thief in sight. But false positives can be horrific, too: During the U.S. military operations in Afghanistan in 2002, the military sometimes detected gunfire, assumed they were under attack, and retaliated. Sometimes the gunfire was not hostile—once it was a wedding celebration and another time it was the Canadians, and many innocents died as a result. Because detection systems, like all other systems, are prone to failure, defenders must remember that no detection system is ever perfectly, permanently accurate. They will sometimes fail to detect an attack and sometimes detect an attack that is not there.

Earlier I talked about vulnerabilities caused by the rarity of events. Detection systems frequently suffer from rarity-based failures. No matter how much training they get, airport screeners routinely miss guns and knives packed in carry-on luggage. In part, that's the result of human beings having developed the evolutionary survival skill of pattern matching: the ability to pick out patterns from masses of random visual data. Is that a ripe fruit on that tree? Is that a lion stalking quietly through the grass? We are so good at this that we see patterns in anything, even if they're not really there: faces in inkblots, images in clouds, and trends in graphs of random data. Generating false positives helped us stay alive; maybe that wasn't a lion that your ancestor saw, but it was better to be safe than sorry. Unfortunately, that survival skill also has a failure mode. As talented as we are at detecting patterns in random data, we are equally terrible at detecting exceptions in uniform data. The quality-control inspector at Spacely Sprockets, staring at a production line filled with identical sprockets looking for the one that is different, can't do it. The brain quickly concludes that all the sprockets are the same, so there's no point paying attention. Each new sprocket confirms the pattern. By the time an anomalous sprocket rolls off the assembly line, the brain simply doesn't notice it. This psychological problem has been identified in inspectors of all kinds; people can't remain alert to rare events, so they slip by.

The tendency for humans to view similar items as identical makes it clear why airport X-ray screening is so difficult. Weapons in baggage are rare, and the people studying the X-rays simply lose the ability to see the gun or knife. (And, at least before 9/11, there was enormous pressure to keep the lines moving rather than double-check bags.) Steps have been put in place to try to deal with this problem: requiring the X-ray screeners to take frequent breaks, artificially imposing the image of a weapon onto a normal bag in the screening system as a test, slipping a bag with a weapon into the system so that screeners learn it can happen and must expect it. Unfortunately, the results have not been very good.

This is an area where the eventual solution will be a combination of machine and human intelligence. Machines excel at detecting exceptions in uniform data, so it makes sense to have them do the boring repetitive tasks, eliminating many, many bags while having a human sort out the final details. Think about the sprocket quality-control inspector: If he sees 10,000 negatives, he's going to stop seeing the positives. But if an automatic system shows him only 100 negatives for every positive, there's a greater chance he'll see them.

And it's not just people whose perceptions suffer when faced with numbingly uniform sensory input. During the World Trade Center rescue operation, the search dogs appeared to be depressed about not finding any bodies, live or dead, so some rescue personnel hid in the rubble for the dogs to "find." Why? Because a depressed dog is inattentive and won't be paying attention when its "play" is for real.

All detection systems have weaknesses, and an attacker who can probe, or test, them without consequence can learn the best ways to circumvent them. For example, pickpocketing is a skill. There are some people who are expert cutpurses, but for most people it's just too dangerous to probe people's defenses and learn the skill. The victim's natural detection ability ("Hey! He tried to steal my wallet!") means that a rookie pickpocket, still low on the learning curve, will pay a steep price. In *Oliver Twist*, Fagin ran a pickpocket school where the children would practice on each other, but in general becoming a better pickpocket isn't easy because it is hard to probe pickpocking defenses. Compare this situation to the world in Larry Niven's "Known Space" stories, where pickpocketing is legal. People can practice the skill as they would any other one—undersea diving, public speaking, bocce—and get better, because the consequences for a mistake are much less severe. But in the real world, the security systems that defend against pickpockets are resilient against being probed. Unsuccessful or clumsy probes are punished, which limits them.

Probes are, of course, a standard tool in spying. During the Cold War, U.S. planes regularly penetrated Soviet airspace in attempts to probe that country's military detection capabilities. These attempts included high-altitude probes by U2 planes (Gary Powers was flying at 68,000 feet when the Soviets shot him down in 1960), as well as lower-altitude flights designed to cause Soviet defenses to turn on to analyze their defenses. The U.S. Air Force plane that collided with a Chinese plane and then was forced to land in China in 2001 was doing the same thing.

Some probes, however, can be discouraged with less spectacular countermeasures. Consider the PIN that protects your ATM card. If a criminal steals your debit card, she can't use it without the PIN. Most PINs are four-digit numbers; there are only 10,000 possible PINs. Theoretically, a criminal could stand in front of an ATM and try them all, one after another. Assuming she could try four per minute, it would take at least 42 hours, which might be manageable for someone desperate. But in fact the system is secure because the ATM network detects this kind of probing and responds to it. If you type an incorrect PIN into an ATM several times in a row—the exact number varies—the network will disable your account, keep the card, or both. Compare this to a similar system that doesn't have any probe detection: a combination padlock. An attacker can stand in front of a padlock and try combination after combination until he finds the right one. A four-digit combination lock is no match for a bored teenager.

Many computer systems have been built, and continue to be built, without any ability to detect password guessing. They let users test as many bad passwords as they want. This lack of detection mechanisms, combined with the ability to automate the attack, means the difference between a secure system that has only 10,000 possible PINs and an insecure system that has billions of possible passwords.

. . . .

Audits are another important security mechanism. An audit is, simply, the practice of reviewing data in an effort to catch attacks (or mistakes). It is an after-the-fact detection system, often cheaper than prevention or immediate detection. Obviously, it works best in situations where it is possible to revisit the event after the fact and then redo it correctly—it is more effective against fraud than against murder—but when it can work, it is often the cheapest way to provide security.

Audit isn't the only way to secure tax collection. The tax authorities could go to every person individually and manually determine what their tax is; this is how it was done in much of medieval Europe. They could also review every tax form upon submission: a detection/response security system for taxation.

Basically, the tax authorities assume that everyone fills out their tax returns correctly. After the fact, sometimes years later, the auditor verifies selected returns to ensure that the forms were filled out correctly and—more important—that the correct amount was paid. If a return was not done right, the tax authorities redo it correctly. Of course, because this is a random detection system, the civil or criminal penalties when someone is caught must be so great that people will believe attempting fraud isn't worth it in the long run.

Double-entry bookkeeping, an important auditing tool, was codified in 1494 by Luca Pacioli of Borgo San Sepolcro, although the concept may be as much as 200 years older. The basic idea is that every transaction affects two or more accounts; one account is debited by an amount exactly equal to what the other is credited. All transactions are transfers between pairs of accounts, and the sum total of all accounts is always zero.

This system has two main purposes. Different clerks keep the two books, reducing the possibility of fraud. (As I said, one of the ways to increase security in a system with trusted people is to have them watch each other.) But more important, the two books are routinely balanced against each other, so that a discrepancy in one book is noticed because it won't balance with the other. (Businesses balance their books every month; banks, every day.) This balancing process is in effect an audit: If one clerk tries to commit fraud—or simply makes a mistake—the balancing process will catch it, because someone else is checking the results. Of course the question arises: What if both clerks are in collusion? Standard practice is for additional outside auditors to come in regularly and check the books just to make sure—in other words, even more defense in depth.

The banking system uses audits all the time: to help secure internal systems and to help secure consumer systems. In addition to double-entry bookkeeping, banks have complex and comprehensive audit requirements. Bank employees above a certain level must take at least one vacation annually of two contiguous weeks, so that their substitutes can detect any possible fraud while they are gone.

Prisons use auditing, as do nuclear missile silos and retail stores. A prison might keep a record of everyone who goes in and out the doors, and balances the record regularly to make sure that no one unexpectedly left or unexpectedly stayed. (Even so, suspected Basque terrorist Ismael Berasategui Escudero escaped from a high-security French prison in August 2002 by switching places with his brother, who was visiting him. I don't know how the two managed to thwart one of the audit mechanisms: The visitors' hands are stamped with an invisible ink before the visit, and that ink is checked after the visit.)

In a missile silo, security officials might go even further and audit every box and package that enters and leaves, comparing shipping and receiving records with another record of what was expected. A retail store keeps a register tape of all transactions and compares how much money the register says is in the drawer with what is actually in the drawer.

Most surveillance cameras are used for audits, rather than real-time detection. Nobody watches the cameras attached to ATMs to detect attacks. If an attack occurs, then the police pull the video recordings. Major department stores may have one or two people checking a camera, but there are far more cameras than guards. Even in casinos, which have cameras trained at every gaming table and most gaming machines, few people are actually monitoring them. The cameras are installed so as to compile a record on tape. If there is a problem, casino security can review the appropriate tape. Even in a building where a guard has a few monitors at his station, the primary value of the cameras is audit. And the concept of auditing is at the heart of gun registration and ballistics tests, as are the various fingerprint (and DNA) databases around the world.

To repeat, auditing is primarily retrospective, after the fact. But it can act as a preventive measure. Most attackers want to get away with their attack, and if they know they'll be caught in an investigation, they're less likely to attack in the first place. A valet might be in a perfect position to steal a car, but she knows she's going to be a prime suspect, camera or no camera.

A system's particular needs for audit depend, of course, on what the security system is protecting. You don't need much of an audit trail for a stored-value card system on photocopy machines at a university; you need a much stronger audit trail if the cards are going to be used to make high-value purchases that can be converted back to cash. When you cash a check of less than $30,000, neither your bank nor the issuing bank bothers to check the signature. Assuming there's enough money in the check's account, you'll get the money. Banks realize that because check fraud is relatively rare, it is much cheaper to get stuck with the occasional loss than to prevent it in the first place. In the U.S., approximately one check in a thousand is fraudulent. If there's fraud involved—say, the signature on the check was forged— the bank expects that to be caught in an audit. The bank expects checking account owners to balance their checkbook at the end of the month, notice a debit they didn't authorize, and then call the bank. The bank will then unravel the transaction as best it can and recredit the money to the checking account owners.

Credit card companies do the same thing. They initially assume that all transactions are good. If there's a fraudulent transaction, they expect customers to notice it on their bill and complain. The company will then credit the customers and, if it wants to, investigate the fraud.

Compare these security solutions to the one banks use to defend against bounced checks. Because bounced checks are much more common than forged checks, and computers can check balances automatically, banks have decided that it is cheaper to defend against this attack by preventing it rather than auditing it afterward—especially since they can assess fees against the account holder.

Not much security surrounds prescription drugs. There's a lot of secrecy—nondoctors don't generally know how to write a prescription—but anyone with the right knowledge, an official-looking prescription pad, and sloppy-enough handwriting can produce a prescription that will fool a pharmacist. For controlled substances, security is primarily based on audit. Doctors are required to include their DEA registration number on prescriptions. The prescription information is entered into a central DEA computer, which can be audited for signs of fraud. Some states even require controlled-substance prescriptions to be filled out in triplicate, with copies going to an enforcement agency. If a forged prescription is identified, it is usually because the pharmacist got suspicious and called the doctor.

The New York Stock Exchange, like other exchanges, constantly monitors trading to detect aberrant activity that may indicate illegal trading practices. Aberrant activity is not just insider trading, but also violations of auction market procedures, instances when price changes in stocks seem too wide, frontrunning (when a firm trades a stock in advance of issuing a research report), books and records violations, and other violations of rules governing members' on-floor trading. Computers flag stock movements and trades that seem suspicious, and then investigators look into the details.

Anonymous systems, by their nature, don't allow a complete audit. Audits work because there is an audit trail that can be reviewed in the event of a problem. Anonymity requires some of that audit trail to never be created. If you walk into a store and make a cash purchase, and then get home and notice that you don't have the right amount of change—or the merchant notices that the cash register doesn't balance at the end of the day—there's no way to go back and figure out what happened. The anonymous transaction is therefore unauditable. I suppose you could audit the transaction if you had a surveillance camera trained on the cash register, but then the transaction would no longer be anonymous.

After the 2000 U.S. presidential election, various pundits made comments like: "If we can protect multibillion-dollar e-commerce transactions on the Internet, we can certainly protect elections." This statement is emphatically wrong—we don't protect commerce by preventing fraud, we protect commerce by auditing it. The secrecy of the vote makes auditing impossible. Another difference between large financial transactions and voting is that in voting, appearances matter. If someone claims to have stolen a billion dollars and no one seems to have lost it, you can safely ignore his boast. On the other hand, if some political group claims, on election night, to have hacked the vote ... what do you do? You can't disprove the claim. You can't redo the vote to make sure it is accurate. And if the people don't believe that the election was fair, it might as well not have been.

Computers can make auditing difficult. Register tapes make good audit records because clerks cannot change them. Transactions are printed sequentially on a single sheet of paper, and it is impossible to add or delete a transaction without raising some suspicion. (There are some attacks on register tapes: blocking the writing, simulating running out of ink, disabling the writing for a single transaction, forging an entire tape.) Computer files, however, can be erased or modified easily, which makes the job of verifying audit records more difficult. And most system designers don't think about audit when building their systems. Recall the built-in audit property of double-entry bookkeeping and the natural resilience of having two trusted people watching each other. That auditability fails when both books are stored on the same computer system, with the same person having access to both. But most computer bookkeeping programs work in this way.

. . . .

Prediction is another important detection technique; it's advance detection. If you can detect an attack before it actually occurs, you can respond to it faster. Perhaps you can even stop it before it occurs, rather than responding to it while it is ongoing or after it has occurred. We all do this; we try to detect what we perceive to be insecure situations, and respond to them by taking precautions.

We do it when we avoid a dangerous neighborhood while walking from one place to another. Companies do it after they have a massive and unpleasant layoff, hiring extra guards as a precaution against workplace violence. The police do it when they suspect a terrorist attack against a particular building, and increase police presence in the area. My mother was capable of advance detection: "Don't even think about it!"

Advance detection decisions are based on two things: information about the environment as it is happening and information about previous similar situations. Increasing police protection around a potential terrorist target is an example of the former; it's done because of what's called a current credible threat. Bringing in guards after a layoff is an example of the latter; companies know that workplace violence after a layoff is not uncommon. Crossing the street to avoid the stranger is a combination of the two: current information about the person approaching and the deserted street (an example of profiling), and previous information about the horrible things that scary-looking people have done or been reported to do. Formalized profiling systems are another example of advance detection: making decisions about a specific person based on demographic information about similar people.

As you might expect, prediction works well when you're right and less well when you're wrong. And since it's prospective (and not retrospective, like an audit), it has a fundamental problem: The trade-offs are extreme because it's necessarily inefficient. We pay the cost every time we take a precaution, but receive a benefit only in the event of an attack. But attackers are rare, and most precautions are wasted. The U.S. has spent millions posting National Guard troops around airports, national monuments, and other potential terrorist targets, and it is doubtful if the minimal additional security was worth the expense. Many children are taught never to talk to strangers, an extreme precaution with minimal security benefit. But as with everything else in security, it's a trade-off, and a particularly subjective one at that.

Predictions can be based on information ranging from specific to general, and the efficacy of the prediction is often a function of the specificity of the information. A mother might tell her children not to talk to strangers as a general principle, or she might tell them not to talk to strangers because a registered sex offender just moved into the neighborhood. Precautions based on specific information are more likely to be worth the trade-offs; predictions based on general principles are more likely to be based on superstition or stereotypes.

When governments conduct organized advance detection, it's called intelligence gathering. Historically, intelligence was primarily a problem of data collection. A country had to send a spy to a foreign country to look around, see what was going on, and report. With the advent of modern surveillance technologies—spy satellites, remote drones, electronic eavesdropping stations—the problem has shifted to data analysis.

There's a world of difference between data and information. In what I am sure was the mother of all investigations, the CIA, NSA, and FBI uncovered all sorts of data from their files, data that clearly indicated an attack was being planned for 11 September 2001. For example, the NSA intercepted two messages on 10 September 2001: "The match is about to begin." "Tomorrow is zero hour." Neither was translated until 12 September 2001.

With the clarity of hindsight, it's easy to look at all the data and point to what's important and relevant. It's even easy to take all that important and relevant data and turn it into information, drawing lines from people in flight schools, to secret meetings in foreign countries, to relevant tips from foreign governments, to immigration records, apartment rental agreements, phone logs, and credit card statements. It's much harder to do before the fact, since most data is irrelevant and most leads are false. In hindsight, it's uncanny to see the disparate pieces of evidence and intelligence that pointed to the 9/11 attacks. But at the time, the crucial bits of data are just random clues among thousands of other random clues, almost all of which turn out to be false or misleading or irrelevant.

The U.S. government has tried to address it by demanding (and largely receiving) new powers of surveillance and data collection. This completely misses the point. The problem isn't *obtaining* data, it's deciding which data is worth analyzing and then interpreting it. So much data is collected—organizations like the NSA suck up an almost unimaginable quantity of electronic communications, the FBI gets innumerable leads and tips, and U.S. allies pass along all sorts of information—that intelligence organizations can't possibly analyze it all. Deciding what to look at can be an impossible task, so substantial amounts of good intelligence goes unread and unanalyzed. Data collection is easy; analysis is difficult.

All criminal investigations examine audit records. The lowest-tech version of this is questioning witnesses. During the 9/11 investigation, the FBI examined airport videotapes, airline passenger records, flight school class records, financial records, and so on. And because it did such a meticulous job examining these records, the investigation was effective and the results were thorough.

We all leave a wide audit trail as we go through life (credit cards, phone records, security cameras, and so on), and law enforcement can already access those records with search warrants and subpoenas. FBI agents quickly pieced together the identities of the 9/11 terrorists, and the last few years of their lives, once they knew where to look. If the agents had thrown up their hands and said "We can't figure out who did it or how," they might have had a case for needing more surveillance data. But the agents didn't, and the FBI doesn't.

More data can even be counterproductive; with more data, you have the same number of "needles" and a much larger "haystack" to find them in. The NSA and the CIA have been criticized for relying too much on electronic surveillance and not enough on human intelligence. The East German police collected data on 4 million East Germans, roughly a quarter of their population. Yet even they did not foresee the peaceful overthrow of the Communist government; they invested too heavily in data collection while neglecting data interpretation. We need more intelligence agents squatting on the ground in the Middle East arguing about the Koran, and fewer sitting in Washington arguing for changes in the law.

The advent of intelligence gathering from public sources, called "open source intelligence," reinforces the point. There are organizations that collect intelligence without resorting to spying; they read and analyze public documents, translating diverse data into coherent information. The data is out there for the taking—if only someone were able to analyze it.

Primarily, advance detection allows defenders to add more security countermeasures. If the government knows of a credible terrorist threat against a particular monument, it can post more guards until that threat subsides. Some companies provide temporary security to a company after a bomb threat; this is another good example of advance detection resulting in more prevention.

A secondary value is that advance detection allows defenders to prioritize security resources. By relying on profiling to decide whom to stop at national borders or whom to search at airport security checkpoints, security agents try to make the most efficient use of their time. By posting guards at national monuments only when intelligence services believe there is a credible terrorist threat, instead of all the time, we save an enormous amount of money.

Two last points. Before criticizing intelligence failures, remember that we rarely have enough context to judge intelligence efforts. In the intelligence world, successes remain out of the news. How many terrorist attempts have been thwarted in the past year? How many groups are being tracked? We have no idea. If a government intelligence agency succeeds in stopping a terrorist attack, an assassination attempt, or a coup, no one ever knows about it. The agencies want it that way, to protect their methods and sources. It's only in failure that they get any recognition.

Also, the problem with advance detection is that it's often wrong and almost always not provably right. When you arrest someone for planning a crime, you really don't know if he was actually going to go through with it. And you certainly can't prove intent. Relying on advance detection to make arrests quickly drifts into the realm of thought crimes, which flies directly in the face of a free society and is more suitable to totalitarian regimes.

Example: CAPPS

In 2002, Al Gore was randomly stopped and searched twice in one week while trying to board an airplane. Would a putative terrorist watching this think (1) Wow, those Americans are so conscientious about terrorism that they search Al Gore? or (2) Wow, those Americans are so stupid that they waste their time searching Al Gore while ignoring people like me? Or would he think (3) Wow, what has Al Gore done—if the system flags him, then he must be guilty of something?

CAPPS, the Computer-Assisted Passenger Profiling System, is a profiling system deployed by the FAA since 1999 to identify high-risk individuals for increased security attention. (The current version is CAPPS-II.) CAPPS rates all passengers using approximately forty pieces of data. The details are secret—the Department of Justice insists that ethnicity is not one of the criteria—but are believed to be based on: the traveler's address, credit history, police and tax records; flight origin and destination; whether the ticket was purchased by cash, check, or credit card; whether the ticket is one way or round trip; whether the traveler is alone or with a larger party; how frequently the traveler flies; and how long before departure the ticket was purchased. This all sounds like a good idea: Why waste precious time making Grandma Lillie from Brooklyn empty her purse, when you can search the carry-on luggage of Anwar, a twenty-six-year-old who arrived last month from Egypt and is traveling without luggage?

Step 1: What assets are you trying to protect? Airplanes, airlines, the passenger air traffic system, and the nation from terrorism.

Step 2: What are the risks to those assets? Terrorism. Specifically, those threats that originate from a hostile passenger boarding a plane. CAPPS won't have any effect on service personnel or antiaircraft rockets.

Step 3: How well does the security solution mitigate the risk? CAPPS is a cross between a profiling system and a random detection system. If the government gets its rating system right, it does some good. But CAPPS is a brittle security system. Its effectiveness is partly a result of the secrecy of the CAPPS profiling criteria.

On the other hand, a smart terrorist can bypass this system pretty easily. Assume a terrorist gets a fake ID and a credit card issued in that name. He buys a ticket using that card; for extra verisimilitude he can buy a first-class ticket and wear a suit. The only metric the CAPPS system can use to pick him out of the crowd is that he's a first-time flyer.

There are two basic problems with the system. First, CAPPS can be probed. If Anwar is searched and isn't found to be carrying anything, he's allowed to board the plane. This means that a terrorist organization can probe the system repeatedly and identify which of its people, or which ticket characteristics, are less likely to be stopped. Against a sufficiently diverse terrorist group, this system is less secure than random searching because the terrorist group can choose to send those of its members whom CAPPS always passes. And second, a system that is more likely to select a certain class of passenger is also less likely to select other classes of passengers. If a terrorist happens not to match the profile, he is going to slip by unnoticed.

Step 4: What other risks does the security solution cause? The database is a significant risk. The database associated with CAPPS is enormous, and data will be stored there forever. There's no way for people to verify the accuracy of information in the database or to correct errors. For the system to be effective, this database must be accessible to a large number of people, any of whom could use it to violate privacy or, even worse, deliberately add false information.

Also, automated profiling systems are a poor substitute for due process. Replacing the traditional system—presumption of innocence and the right to face your accuser and defend yourself against accusations—with a mathematical algorithm is very risky; due process laws exist for a reason.

Step 5: What trade-offs does the security solution require? Money, time, infrastructure: the usual trade-offs we see in every security countermeasure. Also the severe loss of dignity, privacy, and personal liberty for those who are innocent but fit the profile. And don't forget the complete inability to fly for people on the named no-fly list who are listed incorrectly.

In the end, a system like this just might be worth the trade-offs. Systems like CAPPS will be likely to single out lone terrorists who meet the profile, or a terrorist group whose entire membership fits the profile, or copycat terrorist groups without the wisdom, time, and resources to probe. But terrorists are a surprisingly diverse lot: Shoe-bomber Richard Reid is British, with an English mother and a Jamaican father. José Padilla, apprehended in May 2002 for plotting to release a dirty bomb (although it seems that the charges were severely overstated), is a U.S. citizen of Puerto Rican origin. CAPPS works best combined with random screening; defense in depth again. It's not perfect, but it's probably the best way to balance the different risks.

Example: Credit Card Profiling Systems

People who steal credit cards follow a common usage pattern; the criminal uses the card to buy a large number of expensive items quickly, commonly choosing items that can be returned or fenced for cash. If the credit card company can detect this spending pattern quickly—if possible, even before the account holder notices that her card is missing—then the credit card company can disable the card and minimize its losses. To this end, engineers have developed artificial intelligence–based antifraud profiling systems that continually sift through charge data in real time, detecting suspicious spending patterns. Sometimes the patterns are so obvious that the company immediately disables the card and then waits for the account holder to complain (if, in fact, disabling the card was an error). Sometimes the company calls the cardholder and asks if the charges are legitimate. If the theft was not of the physical card, but simply the account number, the alternative would be to wait until the monthly statement arrives and the account holder notices the erroneous charges. We can examine the security system using our five-step process:

Step 1: What assets are you trying to protect? Money. The asset is not the physical card or the physical number. The asset is really the attacker's ability to use someone else's credit card account to buy merchandise fraudulently. The mechanism is irrelevant.

Step 2: What are the risks to those assets? The specific risk is that there can be a long delay between the time a credit card (or number) is stolen and the time the card is reported stolen. Credit card thieves know that they need to use cards quickly, before the account is shut down. Reducing that delay saves the credit card company significant money.

Step 3: How well does the security solution mitigate those risks? The details depend on how accurate the fraud-detection system is, but in general it is an excellent solution.

Step 4: What other risks does the security solution cause? It is possible, in theory, for the system's programmers to install specialized code that would hide their own credit card thefts, but that's a minimal risk. A greater risk is that the criteria the system uses become public, and criminals can tailor their fraud patterns so that it remains undetected.

Step 5: What trade-offs does the security solution require? The system costs money to implement and maintain. Another trade-off is user annoyance. If the system gives too many false alarms, users will start ignoring the telephone calls and complaining about the intrusions.

This analysis tells us that the countermeasure is definitely worth it, and that's why all credit card companies have implemented these detection systems. There's no way to prevent credit card theft; the best the companies can do is detect it before it becomes a significant problem. Because this is a computerized profiling system, it won't catch everything. But profiling works here because the system profiles actions and not people; there is a very restricted set of information and some clearly defined characteristics to profile on. The system profiles on recent spending patterns, nothing more. As long as the rate of active failures stays low—I get a call from my credit card company about once a year on a false alarm—this system works.

Chapter 12

Detection Is Useless Without Response

Responses fall into five categories. Reaction, directed against the attacker, defends; mitigation, focused on the assets, limits the effects of the attack; recovery repairs the security system after an attack; forensics is the post-attack analysis of evidence to determine what happened; and counterattack turns the defenders into attackers—to exact revenge, but also to prevent future attacks. Deterrence, which broadly includes all the steps taken to prevent future attacks, is another form of response, as is education.

Detection is useless without response. A burglar alarm that rings and rings, with no one to respond to it, is hardly better than no burglar alarm at all. If an attacker knows a car alarm is going to be ignored, the alarm might as well not be there in the first place. Response is what makes detection a valuable security countermeasure.

Of course, nothing is that simple. In Russia, responding to a car alarm may get you shot, because the thieves would rather shoot you than be interrupted in their theft attempt. Car owners have resorted to automatic response systems, such as a bear trap under the gas pedal. By now, of course, thieves have learned to check before getting into cars. But since detection systems are error-prone, there is always the possibility of an erroneous response. Booby traps, like the bear trap under the car's gas pedal, are illegal in many jurisdictions because they pose a threat to police officers, firefighters, tow-truck operators, and other law-abiding people doing their jobs. Remember the accidental bombings of wedding parties in Afghanistan? Response needs to be tempered based on the accuracy of the detection mechanisms.

Rarity is another problem when it comes to response. When attacks are rare, attackers may try to exploit the fact that certain response mechanisms don't get exercised very much. This is why centralized response services work so well. Even the wealthiest people don't hire their own doctor to sit on their estate, waiting for them to get sick; they know they will get better medical care if their doctor has seen patient after patient, building her experience. It's the same with the police, with the fire department, and with the military. We want the paramedics who respond to our calls to have seen our problem and have solved it, again and again and again. We want what is rare and terrifying to us to be familiar, if not routine, to them. Centralized security responses may seem to provide less security because the response is less personalized, and it's not on-site, but in fact they often provide more security because the response is by trained and practiced experts.

There are five different types of response:

- *Reaction* is response directed against the attackers. It's the guards who come running to engage the attackers, when they blow a hole in the castle wall.
- *Mitigation* is response directed toward whatever is being defended. It's the other guards, who spirit the princess away to safety.
- *Recovery* is response after the fact: the guards who rebuild the hole the attackers left in the castle wall, and the engineers who try to figure out how to make the new wall bomb-proof.
- *Forensics* is also after the fact; it's the guards who carefully scrutinize the dead queen's body, trying to figure out how it happened.
- *Counterattack* includes the guards who ride from the castle to exact revenge on the attackers' king.

In this chapter we'll focus on all these forms of response, as well as more general responses—such as deterrence and education—that are not linked to a particular attack but instead improve the general climate of security.

. . . .

The most obvious *reaction* to an attack is to invoke more countermeasures. Doing this works because attacks are rarely instantaneous; more often, they involve multiple steps. Sometimes the best defense is to allow attackers to succeed a little bit, commit themselves, and only then to employ additional defenses. Think of a military compound with fences, motion sensors, and perimeter guards. What generally happens is that an attacker climbs the fence and trips the motion sensors. This alerts the guards, who react to the attack and repel the attacker, even though he has already reached the other side of the fence. Permitting the attacker to climb the fence is not catastrophic, because it's just his first step on his way to achieving his objective.

After scaling the fence, he might then have to break into one of the buildings, then break into one of the offices, and so on. Defenders make use of the time lag between the initial attack and the attacker achieving his objective. Certainly it would be more secure to have a guard permanently stationed every ten feet along the fence, but that's probably too expensive and, in all likelihood, unnecessary. An adaptive defense that detects and responds to the attacker's first intermediate success, before he manages to do anything else, is a much better trade-off between security and cost.

Consider another example: If an attacker just wants to smash a store window and steal the first things he can grab—as might be the case in a jewelry store during a riot—the guards who respond to the alarm have to be very fast to be effective. Because the attack takes almost no time, instantaneous response is required. On the other hand, in a bank an alarm that silently sounds when a robber breaks a window in a branch office, combined with a bank safe that takes a few hours to open and guards who respond to the alarm from their outpost at the main office, on the other side of town, is likely to work just fine. Watch the sequence: prevention mechanisms such as locks on the doors and bars on the windows, detection mechanism in the alarm, more prevention in the bank vault, and response by the guards. This is why banks don't as a rule post 24-hour guards at every branch office and why jewelry stores take their wares out of the windows and lock them up in a vault at night. The vault gives the defenders time to respond.

Sometimes there's no good response: A quick-response setup is too costly, while a delayed response is simply ineffectual. Consider the situation at the U.S.–Mexican border. Many people try to cross daily (actually, nightly), and it isn't feasible to patrol all of a 2,000-mile border. There is some prevention—the immigration checkpoints, and the fences and other physical barriers—and some detection by both patrols and automatic systems. And there is minimal response. But there's simply too much border to cross and too few guards to react.

Not all responses are human. Some safes use a kind of aggressive, almost instantaneous automatic reaction. There's a tempered glass panel behind the safe's wall, and if someone tries to drill out the lock, the glass plate shatters, which triggers all the relock pins. This reaction makes the safe much harder to open, requiring a blowtorch and time. The response countermeasure serves to make a low-end home safe more secure without the need for a more expensive primary prevention countermeasure (a better lock). Note the security trade-off here: added security at the cost of additional repairs to the damaged safe.

Reactive countermeasures often boil down to a series of moves and countermoves between attacker and defender. Consider a group of commandos trying to disable a satellite receiver at a military base. The attackers may try to cause a diversion, attempting to get the defenders to react to the wrong thing. The defenders might send some of their forces to deal with the diversion, but keep the others back in case something else happens. The attackers may attack the reaction mechanism separately—the defenders who respond to the diversion—as well as the original target. The attackers may attack again and again, in an effort to tire out the reaction mechanism. Or they might give up and call in an air strike. Reactions can be subtle and complicated, especially when they involve people. Automatic reactions tend to have only a few tricks, tricks that attackers can learn and then learn to overcome.

. . . .

As mentioned earlier, *mitigation* is that part of an overall response aimed not at the attacker, but at the defender and the assets. It's the automatic sprinkler system that turns on when the smoke detector detects a fire. It's the evacuation procedures that protect the President of the U.S., and the 25-ton blast doors that close the entrance to NORAD in Cheyenne Mountain in the event of a nuclear attack. It's the automatic system that deactivates credit cards or cell phones if fraud is detected, so that further losses cannot accrue. It's the portion of response that assumes failure and attempts to minimize the damage.

Attack mitigation is common in the animal kingdom. An iguana can detach its tail if a predator bites it. The system fails securely; the iguana has suffered some damage, but it has survived and can grow a new tail. Similarly, some sea cucumbers are able to evert their stomachs and still survive. The attacker eats the stomach and its contents instead of the sea cucumber, which can grow another—another successful attack mitigation.

Firefighters use mitigation techniques to defend against forest fires; they dig trenches and start backfires in an effort to limit the spread of the fire they're fighting. A computer network under attack can mitigate the damage by disconnecting from the Internet. Run-flat tires allow a car to keep getting away even if its tires are shot. A sump pump protects a house from basement flooding when the foundation leaks, demonstrating that a quickly mitigated security failure can be just as secure as an unbreakable countermeasure.

There's obviously overlap between reaction and mitigation. Consider an automatic door that closes and locks to protect particularly valuable paintings in the event a museum detects an intrusion. It is both a reaction—it adds a preventive countermeasure in response to an attack—and a mitigation: Even though the robbers may be able to steal some of the museum's paintings, they won't be able to steal the most valuable ones.

. . . .

Recovery ensures that the system survives the attack. It's mitigation, but after the attack is over. It's the backup copy of your critical computer files that you turn to when your hard drive crashes or when someone steals your laptop. Recovery mechanisms allow people in a flooded office building to return to normal, or at least return to work. Sometimes recovery is the easiest way to provide security. To take a trivial example, I don't need to protect my pen against theft. If it's stolen, I use a different one. It's easier for me to recover from the attack than it is to defend the pen.

There are many different ways to recover, depending of course on exactly what you're trying to recover. When your ship hits an iceberg, you can recover the ship by operating the bilge pump, or you can recover the passengers and crew by manning the lifeboats. Which you choose may depend on how bad the hole in the ship's hull is. First, though, you must prepare yourself for recovery by installing both lifeboats and a bilge pump. (Depending on how you look at it, this might be considered mitigation as well.)

The nature of the recovery depends on which aspects of the assets are most important. If the sinking ship is a cruise ship, the most important recovery consideration is to get everyone into lifeboats; saving the vessel is secondary. On the other hand, sailors on a warship will remain on a damaged ship longer, risking their own lives, and work toward keeping the ship afloat and fighting; during wartime, a functional ship is more valuable. Again: trade-offs.

The system of recovery determines the countermeasures. Before the invention of radio telegraphy, the odds of a sinking ship in the middle of the Atlantic being intercepted by another ship were very, very low. Hence, lifeboats were the only recovery tool available, and they had to hold everyone for a long time. After the invention of radio, ships could more or less count on help arriving within a day, reducing both the need for lifeboats and the amount of supplies on them.

Huge centralized corporate computer systems often have recovery procedures in place, but they differ widely from each other, depending on the business need—the particular kind of asset protection—that's important to the company. Credit card companies have massive backup facilities because they need to recover operations as quickly as possible, and they can't afford to lose any data in the process. In contrast, a phone company wants its switches to recover as quickly as possible but doesn't care if everyone has to redial phone calls. Obviously, the more extensive the recovery a system needs, the more expensive and difficult it is to implement, and the more critical it becomes to define recovery goals and policies beforehand. Developing a recovery plan in the middle of an attack doesn't work.

Recovery often means bringing a system back to the way it was before, which means that it is still vulnerable to the type of attack that caused the need for recovery in the first place. If the attack is rare enough—a bank robbery, for example—this might be okay. If the attack is common—an attack against a computer system—or a class break, this might be naïve. Smart recovery includes a reassessment of attacks, attackers, threats, and risks, and probably some enhancement of security.

••••

After-the-fact evidence collection and analysis is called *forensics*. It's figuring out what happened, and how. Forensics can be essential in identifying attackers and amassing sufficient evidence to charge and convict them. It also provides information so that the same attack can be better defended against next time. And it satisfies the human desire for understanding.

Evidentiary requirements make forensics difficult. Courts require the prosecution to prove the evidence is authentic and could not have been tampered with. This requirement imposes a completely new security problem on the practice of evidence collection and preservation. There's a whole lot of science in modern forensics, and not nearly enough space in this book to do the subject justice. Some of the more interesting developments in the field involve new and complicated techniques that can be difficult to explain to a jury, and old and accepted techniques falling out of favor. DNA evidence is *de rigueur* in some types of prosecution, while recent evidence indicates that fingerprints aren't nearly as reliable as criminologists once thought they were. One interesting development is that as forensic evidence gets better, defense attorneys are more likely to try to impeach the process of evidence collection and analysis rather than the evidence itself. They've realized that forensics is no more accurate than the trusted people who perform the forensics, and that those trusted people are the weak link.

Forensics and recovery are almost always in opposition. After a crime, you can either clean up the mess and get back to normal, or you can preserve the crime scene for evidence. You can't do both. Forensics is often bad for victims in the short term, because they want to get back to normal as quickly as possible. However, forensics can be good for everyone else, because catching attackers is better for society. How this trade-off is resolved in a specific case depends on the power of the various players involved. It's a common dilemma with respect to computer attacks. In their rush to return the network to its fully operational state, system administrators regularly destroy evidence that can be used to convict hackers. The same problem exists with highway accidents: It's often more important to clear the wrecked cars and restore traffic flow than it is to keep the evidence pristine for forensic purposes.

. . . .

Counterattack means turning the tables and attacking the attacker. It's less about fighting back and more about attacking in retaliation. Against a military air assault, it could involve flying planes and attacking the enemy's airfields, fuel depots, and ammunition storage facilities, but it could also mean attacking the enemy's military headquarters. Note that the line between defense and offense can blur, as some counterattack targets are less clearly associated with a specific attack and are more geared toward denying the attacker the ability to wage war in general. Legal prosecution is, in some sense, a form of counterattack.

The history of warfare teaches again and again that counterattack is a very effective form of defense. The Battle of Chancellorsville in the American Civil War, the Battle of the Bulge in World War II, and the Six-Day War in 1967 were all won through judicious use of counterattack. The tactics of Sun Tzu, Napoleon, and Rommel all stress the importance of counterattack in military campaigns. Throughout history, intelligent counterattack has worked.

It works in the animal kingdom, too. The woodland ant (*Pheidole dentata*) survives in areas dominated by fire ant colonies, even though the fire ant colonies both tend to be up to one hundred times larger and make lousy neighbors. The woodland ant's trick is counterattack: It has a permanent population—about 10 percent—of warrior caste ants that do nothing but patrol with the worker ants. Whenever any of these warrior ants sees a fire ant, they rush it, get some of its smell on themselves, and then run home, laying a scent trail. They immediately alert the hive, and any other workers and warriors they encounter along the way. A huge force of woodland ants arrives shortly, kills the offending fire ant, and then spreads out to search for survivors. The warriors will keep circling the area for hours, looking for and killing any other fire ants they find. By throwing the maximum possible effort into this counterattack, the woodland ants ensure that no fire ants get back to their own hive with accurate information about the woodland ants' whereabouts.

The British did a similar thing with German spies during World War II by turning every single German spy in Britain into a counter-spy. They were all offered the option of secretly working for the British or dying. Because there were no loyal German spies in Britain, accurate information never got back to the German High Command, and the counterspy ruse was successful. In both cases, this system is brittle; it must be perfect to succeed. A single fire ant returning to the hive will cause its brethren to swarm and possibly wipe out the wood-land ants (although they have some other defensive tricks, including grabbing their eggs and abandoning the nest); a single undiscovered German spy could expose all the counterspies (although presumably the British had some other tricks, too).

While defense in American football is mostly about detection and response, there is the occasional counterattack. A blitz is a defensive counterattack against the offensive quarterback: The linebackers, who usually hang back to defend against a short pass or a runner breaking through the line, instead charge forward and attempt to flatten the quarterback. Because of limited resources, the defense needs to decide if it will devote those resources to basic defense—detection of oppo-nent actions and response to them—or counterattack; it cannot do both. If the counterattack works, this is a great play. If it doesn't, the attacker usually has an overwhelming advantage over the defender and can gain a lot of yards on the play.

The key difference between warfare and many of the other exam-ples in this book is the variety of options available to the defender. After being attacked, a general can turn around and counterattack the attacker—something a bank cannot do against a perpetrator of credit card fraud. If you catch someone burglarizing your home, in most countries it is illegal for you to follow him home and steal your stuff back. Reaction is often legal—you can defend against the burglar in the process of robbing your home—but you can't retaliate. Counterattack needs to be done *very* carefully.

There is a notion of a counterattack based on a prediction: a preemptive attack. This is very difficult to justify as a defense, both because predictions are so error-prone and because there is not yet any attack to justify a response. You can't jail someone for shoplifting before she actually shoplifts, for example. Overthrowing a government before they attack you is itself an attack; it is not a defense. This does not mean that a preemptive attack cannot be justified—many justified the U.S. invasion of Iraq as a preemptive attack against an unstable and unpredictable regime trying to obtain nuclear weapons—but that it cannot be justified solely as a defensive response.

In warfare, the lines between attack and defense blur, and the defender has much more flexibility as a result. There's an old saying that the difference between an offensive weapon and a defensive weapon is the direction it's pointing. The difference between offense and defense is often more a matter of political maneuvering and public opinion than anything else.

. . . .

Prisons exist for any combination of four reasons: revenge, removal, deterrence, and rehabilitation. Only the latter three—removal, deterrence, and rehabilitation—have any implications for security. Throughout history, rulers have thrown people in jail to remove them from society or take them out of politics. Long prison terms for child molesters have been justified because they keep the irremediably guilty off the streets so they can't repeat their crimes. The ancient Greeks invented the concept of ostracism, where people would be banished from Athens for a period of ten years. Sir Walter Raleigh was accused of treason by King James I and imprisoned in very elegant quarters within the Tower of London for thirteen years. Both Stalin and Mao jailed, and killed, millions to shut them up. More recently, the Burmese government kept Aung San Suu Kyi under house arrest for years in an attempt to silence her and cripple her political party, and the South African government imprisoned Nelson Mandela for over twenty years.

Deterrence is how security systems prevent future attacks. It's a myth that our society prevents crime. It doesn't. It can't. If someone wanted to kill you, he would just kill you. What we do is detect crime after the fact ("Oh, look; there's a dead body on the floor over there"), collect enough evidence to convince a neutral third party (judge or jury) of the guilty party's guilt, and then punish him. Assuming the culprits are identified and brought to justice, the entire process acts as a preventive countermeasure to deter future crimes.

Deterrence works. In societies without consequences for crime, or where punishments are mild, crime is more common. The more unpleasant the punishment, the better the deterrent. People who are willing to flout the drug laws in the United Kingdom may be unwilling to risk spending time in a Turkish prison for the same offense.

On the other hand, while punishment was much more unpleasant in eighteenth-century England when Dick Turpin was the most notorious highwayman, today there are no more highwaymen in England—but it was the counterattack of finding and capturing criminals that worked, not the severity of the punishment. If attackers think they can evade capture, it doesn't matter what the punishment if captured is. Severe punishment can lead to people thinking: "I might as well be hanged for a sheep as a lamb." Unpleasantness of the punishment is a factor, but not a determining one. Deterrence isn't as easy as it appears.

One of the reasons legal deterrence does work is that laws apply only to extreme behavior. There's no law that prevents someone from calling someone else an idiot. Countries that have laws that regulate normal behavior have a far greater number of people violating those laws.

Deterrence is far less effective against emotional attackers, which is one of the reasons the death penalty doesn't seem to affect the murder rate—the vast majority of murders are emotional crimes. The Israeli government has a policy of bulldozing the homes of the families of suicide bombers. You can argue about the efficacy and morality of this countermeasure, but it is an attempt to deter future suicidal attackers; even if they are willing to throw away their own lives, maybe they'll think twice about destroying the home of their family.

Nonappeasement is another form of deterrence, but it's easier to understand than to put into practice. Everyone knows, for example, that it's a good policy to refuse to negotiate with kidnappers. If you give them the ransom money in exchange for the person who's been kidnapped, then you're just creating an environment where kidnappers know they can make money. Refusing to pay the ransom is much smarter. Even though this might put the kidnapped person in jeopardy, it means that other potential kidnappers will realize that they can't make any money that way. Not paying ransom acts as a deterrent. Nevertheless, watch how quickly someone abandons this policy when a loved one has been kidnapped. It's a classic dichotomy between the agenda of the individual and that of society: It's in the interest of the individual to pay the ransom, but that makes kidnapping a more attractive crime, and society suffers as a result. Human nature is what it is, and personal feelings are what they are; far too often when an executive is kidnapped in a Third World country, his company pays the ransom. In this instance, cold-heartedness provides better security … as long as you can tolerate some victim loss.

The same idea is behind the U.S. government's anti-counterfeiting program. The money the government spends to prevent counterfeiting is more than the value of the counterfeited currency itself. But the problem is that unless counterfeiting is deterred, it will get out of hand, become much more common, and ultimately destroy the credibility and value of the currency. By spending the money, the government is deterring a whole class of potential counterfeiters.

Throughout history, people and countries have used intimidation to deter attackers. The Romans invented the term "decimate," which originally was a description of a procedure: kill one in ten. They would do this to punish rebellious cities or mutinous armies, both as a retribution for the guilty and as a deterrent to others. Organized crime uses the same sorts of tactics to cow people into submission. Someone could use bribery, seduction, or blackmail to achieve similar goals. In the animal world, a deep growl can be enough to scare off a would-be predator.

The Mongols had an interesting tactic to quell opposition. When invading a country (the Mongols are the attackers here, but the tactic works equally well from the other side), they would attack the first city they came to and kill everyone they found. Then they would go to the next city and say: "You can either surrender and submit yourself to Mongol rule, or you can defend yourself and suffer the fate of the previous city." The Mongols were scary enough to get away with this. This kind of approach isn't all ancient history. In 1982, president Hafez al-Assad of Syria leveled the city of Hama, killing between 10,000 and 25,000 of his own people, to demonstrate why opposing his rule was unhealthy. Many governments are not above using similar tactics on the individual level to get prisoners to talk, especially in wartime. Protection rackets operate on the same principle.

....

Deterrence doesn't work unless it goes hand in hand with *education*. For the main part, defense against attacks in progress is not what keeps society safe from crime. Our society is safe from crime because, for the most part, people obey the law. Most people are moral and ethical. Morality is simply human nature, and probably the primary reason our species can have civilization in the first place. Ethics works because we educate our populace as to what is ethical and what being a lawful citizen means.

This is true in the animal kingdom, too. There's a clear hierarchy in a wolf pack, and the individual animals have "laws" they obey. It's also found in primate societies: a clear social order with rules and norms—and violators.

Moral and ethical education is considerably more complicated than what I've written above. Both laws and societal norms impose values, and people are more likely to obey laws that conform with societal norms. For example, both laws and society maintain that stealing is wrong. Hence, stealing is relatively rare. However, there are some cultural groups that maintain that marijuana use is okay, or distributing unlicensed copies of music CDs is okay, so laws prohibiting those behaviors are much more difficult to enforce.

Rehabilitation is a form of education, one designed to take criminals and turn them into law-abiding citizens. Prison without rehabilitation is much less effective, because it prevents crime only as long as the criminals are in jail. In general, deterrence works less well if you don't tell people what they're supposed to do instead. And conversely, education without deterrence works only on those of us who are naturally moral and ethical. (Luckily for our species, this is most of us.) Education is the carrot, and deterrence is the stick. Together, they are a means to excellent security.

Example: Home Burglar Alarms

Should you get a home burglar alarm? Let's find out.

Step 1: What assets are you trying to protect? Your home, its contents, and the people living there.

Step 2: What are the risks to those assets? The risk is burglaries. (Note that this is actually several different risks, depending on whether you're home asleep, home awake, or not home.) There are preventive countermeasures like door locks in place to make it harder for someone to break in, but they're by no means perfect.

Step 3: How well does the security solution mitigate those risks? Home burglar alarms are a detection/response countermeasure. Sometimes the response is simply to make a loud noise, but many alarms are connected to the local police station. False alarms—active failures—are serious problems, however; police in many jurisdictions have stopped responding to burglar alarms, since over 90 percent are false alarms.

Burglar alarms are also a deterrent. A burglar who sees evidence of an alarm system is more likely to go rob the house next door. As far as the local police station is concerned, this doesn't mitigate the risk at all. But for the homeowner, it mitigates the risk just fine.

A skilled and dedicated attacker will most likely be able to defeat your burglar alarm, but most burglars will probably just go rob another house. And burglars also know that any response to an alarm system will take at least a few minutes; they might try to grab things quickly and run.

Step 4: What other risks does the security solution cause? There are minor ones: You have to trust the company that installs the alarm. And there are major ones: The alarm system will have an access code to turn it off; you have to remember it, and you have to give it to everyone who might need it. (Many households never bother changing the access code from the manufacturer's default. Burglars know this.)

Step 5: What trade-offs does the security solution require? Cost and usability are the big ones. A home alarm system isn't cheap, either to install or as an ongoing expense, although newer wireless versions reduce the costs and annoyance of pulling wires throughout your house.

You have to remember to engage the alarm; otherwise it won't work. If you keep the alarm turned on even when you're home, you have to do more work to open the windows. If you turn the alarm on only when you leave the house, you have to remember to close the windows. You may have to adjust the system to be less sensitive to lessen the chance of pets setting off the alarm.

From a strict monetary standpoint, the security solution is probably not worth the trade-offs. But your personal answer is going to be more complicated. If you live in a neighborhood with a high crime rate, a home burglar alarm system is going to be more valuable to you. If you have children who are going to forget to engage the alarm system (or disengage it), the trade-offs are going to be more significant and the alarm won't provide as much security. Perhaps you can get something that only *looks* like an alarm system and save yourself some money.

This decision is a highly personal one, and encompasses more than security. First, there is an emotional cost of attack. Being robbed can be extremely traumatic, especially if (as often happens) there is vandalism, as well. There's a huge difference in emotional impact between paying (say) $1,000 for monitoring services over a period of five years or suffering one break-in that robs you of $1,000 worth of property. And second, there's the emotional benefit to the additional security. If someone in your family is afraid of being robbed and an alarm system will make him or her feel better, then it may be a good trade-off for emotional reasons even though it might not be a good security trade-off.

Chapter 13

Identification, Authentication, and Authorization

The problem with securing assets and their functionality is that, by definition, you don't want to protect them from everybody. It makes no sense to protect assets from their owner, or from other authorized individuals (including the trusted personnel who maintain the security system). In effect, then, all security systems need to allow people in, even as they keep people out. Designing a security system that accurately identifies, authenticates, and authorizes trusted individuals is highly complex and filled with nuance, but critical to security.

It's not sufficient to protect a valuable asset by encasing it in stone or steel, or by sending it to outer space, or by posting armed guards around it. With a very few exceptions, all security barriers need to be penetrated—under authorized circumstances by trusted people. The barrier needs a system that facilitates penetration, and additional systems to determine who is trusted. Buildings and safes have doors and keys or combinations so authorized people can open them. A casino slot machine has a locked door that lets maintenance personnel repair and refill the machine; it also has an opening through which players can collect their winnings—another avenue of penetration, for the user who has been temporarily "authorized" by a winning spin.

The additional security requirements needed to make a barrier conditionally penetrable necessitate an enormous effort of planning, design, and execution: What was once a simple system becomes a complex one. A barrier is designed to keep attackers out; but since we need to allow people in, we must make a barrier that can be penetrated in authorized circumstances and can't be penetrated under any other circumstances. We need to punch a hole in the barrier and then control access to that hole. Our intentionally created holes—windows and doors, for example—are far and away the most frequent avenues for unauthorized entry. The holes we intentionally put in a barrier are very often the weakest link, since they make the security of the barrier depend on the security of the hole and its own systems: identifying the trusted people who are allowed in, the circumstances under which they are to be allowed in, and what privileges they are to have once inside. These ancillary systems of identification, authentication, and authorization are far more complex and subtle than they seem. Understanding the security of barriers means understanding the security of these systems.

In London, you can purchase passes that allow you unlimited travel in the Underground subway system for a week, a month, or a year. Each pass actually has two parts: a photocard and a pass ticket. The photocard is permanent: You get it from a ticket clerk, and it has your picture, an identification number, and a place to write in your name. It has no expiration date and can last for years. It doesn't let you on any trains. For that you also need a ticket, which can be purchased from a vending machine or from a clerk, and which isn't valid unless you write your identification number on it and present it with a photocard.

Many systems jumble identification, authentication, and authorization; in the London Underground system, they're nicely separated. The pass ticket is your authorization. When you enter the Underground system, you either slide the card through a machine that opens a turnstile, or show it to a clerk who's guarding a barrier. (Actually, the barrier might not be staffed—it's a random detection system.) When you want to exit the Underground system, you have to do the same. The ID card is your authentication. The card says: "This photocard is valid for use only by the person shown with a ticket bearing the same number." The number ties the photocard and the ticket together. The photocard is the authentication device, or *token*, but doesn't authorize you. The ticket is the authorization token that allows you access to the Underground.

Notice that this is an anonymous system. There's a place to write your name on the photocard, but you don't have to do it. No one checks IDs to make sure that the name is correct. I could get a dozen photocards in a dozen different names. All the Underground cares about is that two people don't share a weekly, monthly, or annual Underground pass. So it's your face, and the corresponding photo of it on the card, that's the key.

．．．．

Identification, authentication, and authorization. The three concepts are closely related, but in a security system it's critical that we tell them apart. Here's the shorthand guide:

•   Identification: Who are you?

- Authentication: Prove it.
- Authorization: Here is what you are allowed to do.

Conflating the three—running them together, failing to distinguish each from the others—can lead to serious security problems. I'll examine each one in depth, with *authorization* first.

Let's consider what you do when you go to an all-you-can-eat salad bar (an American invention, I'm sorry to say)—and to make matters worse, let's imagine it's at a fast-food restaurant (another American invention; sorry again). Once you've paid for your meal, you're given a plate. With no waiter service, that plate is an authorization token. Because you have one, you are allowed to take salad from the bar. If you brought your own plate, the people working there would (management hopes) notice and stop you, because you did not have your authorization token. (Whether you can break the system by taking a plate home and bringing it back again the next day is left as an exercise for the reader.)

Postage stamps are another authorization token. The goal of the security system is to ensure that senders are authorized to mail a letter; more specifically, that they have paid for the service. A stamp is the authorization token that facilitates that security model. Senders buy these tokens at a post office and, by affixing them to an envelope, transfer that authorization to the piece of mail. This authorization travels with the mail and can be verified by any postal employee at any time. And the token is transferable until it's used; one person can buy it and another can use it.

It's a simple security mechanism. There are attacks—forging a stamp, erasing a cancellation, slipping a piece of mail into the system after the postage is canceled, reusing uncanceled or partially canceled stamps—but they're not easy to execute, and the value of the fraud is low; such a criminal act would make sense only if applied at once to a huge quantity of mail. Stamps work well as authorization tokens.

Tickets—the kind that get you a seat at a theater or a stadium— are also authorization tokens. So are subway tokens: the kind that are good for one ride, the kind that have a balance and are debited with each use, and the kind that are good for a specific trip between two stations. If you've ever walked into a crowded bank or delicatessen and been asked to take a number and wait, the paper with a number on it that you pluck from a machine is an authorization token—it authorizes you to take your rightful place in the service queue.

The London Underground pass ticket is good for a specific period of time within a specific area of travel, but for the most part tokens—theater tickets, postage stamps—are good for one use only. And once the one-time token is used, a trusted person usually does something to the token to render it invalid (like punching a hole in a ticket) or simply takes it away (as in the case of a subway token that's deposited in a turnstile slot). Some tickets are torn in half, postage stamps are canceled, bank checks are marked paid—the one-time authorization has been utilized, and the token is no longer valid.

Authorization tokens don't necessarily require any identification, and generally they are transferable from one person to another. There are, however, important exceptions. Airline tickets (and now long-distance train tickets in some countries), for example, contain passenger identification information (a name) in addition to an authorization to board a specific flight (or train) on a specific date. Compare them with commuter train or bus tickets, which generally don't have any identification information. This system of authorization without identification has even been employed recently by the City of San Francisco in its medical marijuana program: A card that authorizes the holder to receive the drug contains an authenticating picture, but the card is intentionally anonymous.

One of the basic ways to achieve authorization is through *identification*. This is the most important security system our species has, and we implement it widely—I'm tempted to say universally—in our daily lives. You recognize (identify) family, friends, colleagues, and famous people, and once you identify them, you know how much trust you are willing to afford them and what you are willing to let them do. You recognize, and thus identify, your co-worker, so you know she's allowed into her office, although possibly not into yours. The local bartender identifies you as a regular and allows you to run a tab.

In many systems, authorization and identification are so muddled that they're perceived as being the same thing. But knowing who someone is and knowing what he is allowed to do are different. Traditionally, the way we identify ourselves is with a name. In more technological systems, we tend to identify ourselves with a number. Early filing systems and punch-card sorters were designed for numbers, not names, and modern computers have carried on the tradition. A system can be designed so that the numbers are guaranteed to be unique. Names aren't unique, and duplicates can be a problem. Even a name as uncommon as Bruce Schneier has a duplicate. He lives in Illinois, and he's tired of getting my e-mail by mistake. Sometimes these mistakes can get serious. A friend, Ann Harrison, told me a story of sitting on the examining table in a doctor's office, waiting for someone to take a look at a mole she was worried about. When the doctor entered, he had a loaded syringe in his hand and approached her with the full intention of giving her an injection. He said: "Your monthly Botox shot, Ms. Harrison." Ann told him she doesn't get Botox injections, to which the doctor replied: "But you're Ann Harrison, aren't you?" The doctor had two different Ann Harrisons as patients.

This kind of thing happens all the time when names are used as identifiers: People get each other's paychecks and airline tickets, people are falsely arrested and detained. A seventy-year-old woman named Johnnie Thomas is searched every time she flies because John Thomas Christopher was one of the aliases used by Christian Michael Longo, a twenty-eight-year-old man who murdered his wife and children. His name was on an FBI list for three days in January 2002. Similarly, Malcolm Byrd has been arrested repeatedly because a man arrested on drug charges falsely identified himself with Malcolm Byrd's name. In Singapore, some names are so common that the police issue He's-not-the-guy-we're-looking-for documents exonerating innocent people with the same names as wanted criminals.

This is why when you put your ATM card into the slot, the account number identifies you; your name doesn't. The screen may *display* a message like "Hello, John Smith, how can we help you today?" but in fact your card's unique identifier is not your name, but a number. The reasons for this are obvious. You don't want someone else with your name withdrawing your money. Your name isn't even used in the transaction. The only reason it's on the face of the card is so that you know it's yours. In fact, I've seen ATM cards without any name on them. In the bank itself, your account number identifies you. There's a name attached to that account number, just as there is an address, a phone number, and a Social Security number. And if you think about it, the bank doesn't care who you are; it only cares that you're the same individual who deposited the money last week and made those two withdrawals the previous weekend. It only cares if you're authorized to withdraw money. That's the whole point behind numbered Swiss bank accounts; the number is the identification, so no name is required.

Account number, customer number, order number, transaction number: These all serve to identify. Many databases identify people by Social Security number. Driver's licenses and passports have unique numbers on them; that's the real identification information, even more so than the name.

Authentication is another system entirely. When you show your passport to an immigration official, she looks at the name. Perhaps she enters the passport number into a terminal at her desk or scans some encoded information that's printed on the first page. Then she compares the picture in the passport with your face as you stand in front of her. The name and number identify you, and the picture authenticates you.

Two more common examples: Your username on a computer identifies you, and your password authenticates that it's really you. If you have a credit card, the account number identifies you, and the signature authenticates you.

Basically, there are three ways to authenticate someone: by something he knows, by something he has, and by something he is. All these ways have been used from prehistory until the present day, and they all have different security properties and trade-offs.

The first method is "something he knows." Think of passwords, secret handshakes, PIN codes, and combinations to locks. During World War II, American soldiers in Europe would ask strangers cultural questions like "Who won the 1940 World Series?" on the assumption that German soldiers wouldn't know the answer, but every American would.

One of the vulnerabilities of this kind of system is that the verifier learns the secret. Imagine two soldiers meeting in the dark. One asks the other, "What's the password?" The other replies with the password. Now the first soldier knows the password, even if he isn't supposed to know it. The same problem could arise with bank tellers, who could learn the PIN codes of bank customers and then use those codes for nefarious purposes. It is for this reason that many kinds of security systems have computers doing the authentication, not people. Not that computers are infallible, but they're easier to control and less easy to subvert.

The second method to authenticate a person is "something he has." The something might be a physical key, a membership card, or a cell phone SIM card. Like the "something he knows" method, anyone can give this to anyone else. In fact, with either of these methods, all you're really identifying is that the person is of a particular group, not that he is a particular person. Knowing the secret handshake authenticates you as a member of the secret society. Having a copy of a house key authenticates you as one of a group that has access to a given house. I might give you enough information for you to call my bank and withdraw money from my account. When you do this, the bank thinks it is authenticating the account owner, when it is really just making sure that the person on the other end of the phone knows enough information about the account and account owner to be an authorized user of the account.

In ancient Hebrew dialects, the word "shibboleth" means "ear of grain" (or maybe "stream"). According to Judges 12:1–6, the Gileadites defeated the Ephraimites in battle and set up a blockade to slaughter any fleeing Ephraimites. The sentries asked each person to say the word "shibboleth." Any Gileadites stopped were able to pronounce the word with the *sh* sound. The Ephraimites, who had no *sh* sound in their language, were trapped when they pronounced the word with an *s* sound. Depending on your beliefs about accent and language skills, this story is either an example of "something he knows" or the third way of authenticating a person: "something he is."

More specifically, it's an example of "something he has that's a physical part of his body." This is what we normally think of as identification. When we recognize people, we recognize their physical features. On the telephone, we recognize someone's voice. Our ancestors used this type of authentication mechanism even before they evolved into humans. In the animal world, cats spray to mark their territory, dogs sniff each other's butts, and whales have individual songs. More modern versions of this mechanism, called "biometrics," include fingerprinting, voiceprinting, hand geometry, iris and retina scans, and handwritten signatures. Ear shape is a *facial* characteristic that's both reasonably distinctive and hard to alter, although it's not necessarily visible on U.S. passport photos. U.S. green cards and German passports require an oblique headshot, showing an ear. People are generally good at recognizing people by biometrics; machines, less so.

Biometrics have an advantage over passwords and tokens in that they can't be forgotten, although they can be lost. (People can lose fingers in an accident, or temporarily lose their voice due to illness.) And biometrics can't be changed. If someone loses a key or an access code, it's easy to change the lock or combination and regain security. But if someone steals your biometric—perhaps by surreptitiously recording your voice or copying the database with your electronic iris scan—you're stuck. Your iris is your iris, period. The problem is, while a biometric might be a unique identifier, it is not a secret. You leave a fingerprint everywhere you touch, and someone can easily photograph your eye.

Relying on a single authentication technique can be brittle. In the *Odyssey*, Polyphemus the Cyclops captured Odysseus and his men and sealed them in a cave with his sheep. Odysseus poked Polyphemus's single eye out, so when Polyphemus had to let the sheep leave the cave to graze, he could authenticate them only by feel. After watching this process, Odysseus and his men escaped by clinging to the undersides of sheep. Better authentication systems use two or more methods. An ATM, for example, uses "something he has"—an ATM card—and "something he knows"—a PIN. (Then it takes the person's picture, for audit purposes.) A passport is a physical card that is hard to counterfeit and contains a photograph. The door-locking device in my company's office uses both a PIN and a hand-geometry scanner.

Credit cards have two forms of authentication—the physical card and a signature—when used in person, but only one when used over the phone: the information on the card. Credit card companies have tried to improve security by requiring merchants to collect the cardholder's address for card-not-present transactions, but telephone and online credit card fraud is still much greater than in-person fraud (15 to 20 cents per $100, versus 6 cents). Several French banks have recently introduced credit card numbers that are valid only once and are useless if stolen during a transaction, an excellent countermeasure to address the threat. And, for additional authentication, credit cards now have additional digits on the back that are not embossed on the front of the card or on the magnetic stripe.

Many systems perform identification and authentication at the same time. When you recognize a person, you're both identifying and authenticating her. When you look at someone's ID, you are both identifying and authenticating her. Other systems authenticate and authorize at the same time. A door key is an authentication token, and it also opens the door—in effect authorizing entry.

Systems that confuse identification with authentication can have significant insecurities. Again and again I'm asked for the last four digits of my Social Security number as an authentication code, even though my Social Security number is a public identification number. I can't change it. I can't prevent others from having it. It's a unique identifier, but it's hardly a secret: a good number to identify me by, but a terrible one to authenticate me by. Mother's maiden name is a similarly lousy authentication code.

I've described biometrics as an authentication tool, but sometimes they are misused as an identification tool. As authentication systems, biometrics answer a simple question: Does this biometric belong to that person? As a biometric identification system, they must answer the much harder question: Does this biometric belong to anyone in this large database of much-less-reliable biometrics of people? This confusion leads to active failures, and eventually to passive ones.

The reasoning is subtle, so let's work through an example. Automatic face-scanning systems have been proposed for airports and other public gathering places like sports stadiums. The idea is to put cameras at security checkpoints and have automatic face-recognition software continuously scan the crowd for suspected terrorists. When the software identifies a suspect, it alerts the authorities, who swoop down and arrest the miscreant. At the 2001 Super Bowl in Tampa, Florida, cameras were installed, and face-scanning software tried to match the faces of people walking into the stadium with a photo database of people the police wanted to apprehend.

I'll start by creating a wildly optimistic example of the system. Assume that some hypothetical face-scanning software is magically effective (much better than is possible today)—99.9 percent accurate. That is, if someone is a terrorist, there is a 1-in-1,000 chance that the software fails to indicate "terrorist," and if someone is not a terrorist, there is a 1-in-1,000 chance that the software falsely indicates "terrorist." In other words, the defensive-failure rate and the usage-failure rate are both 0.1 percent. Assume additionally that 1 in 10 million stadium attendees, on average, is a known terrorist. (This system won't catch any *unknown* terrorists who are not in the photo database.) Despite the high (99.9 percent) level of accuracy, because of the very small percentage of terrorists in the general population of stadium attendees, the hypothetical system will generate 10,000 false alarms for every one real terrorist. This would translate to 75 false alarms per Tampa Bay football game and one real terrorist every 133 or so games.

That kind of usage-failure rate renders such a system almost worthless. The face-scanning system needs to interact with another system—a security apparatus that must go into high alert with all its attendant cost, inconvenience, disruption, fear, and panic—and will still come up empty-handed in the end. The guards who use this system will rapidly learn that it's always wrong, and that every alarm from the face-scanning system is a false alarm. Eventually they'll just ignore it. When a real terrorist is flagged by the system, they'll be likely to treat it as just another false alarm. This concept, called the "base rate fallacy" in statistics, applies to medical tests, too. Even very

accurate tests can be useless as diagnostic tools if the disease is sufficiently rare among the general population. A 90-percent accurate
system, assuming a 1-in-10-million terrorist density, will sound a million false alarms for every real terrorist. And current systems are much
less accurate than that; in March 2003, an Australian system was
defeated by two Japanese men who simply swapped passports. It's
"The Boy Who Cried Wolf" taken to extremes.

It's not just the face recognition software. The system presumes a
photo database of terrorists. It seems unlikely that terrorists will pose
for crisp, clear photographs. More likely, the photos in the database
are grainy ones taken from 1,000 yards five years ago when the individuals looked different. We have to assume that terrorists will disguise themselves with beards, hats, glasses, and plastic surgery to make
recognition harder. Automatic face-recognition systems fail miserably
under these conditions. And remember, the system I postulated for
this example presumes a face-scanning system orders of magnitude
more accurate than the ones being sold today. A recent test of an airport system indicated it was less than 50 percent accurate, making it
completely useless as an identification system.

Biometric authentication is different. Here the system compares a
biometric on file, called the "reference biometric," with the biometric
of the person at the time it is being compared. This reference biometric is not a blurry photograph taken by an undercover spy; it's a known
clear picture taken under the best lighting conditions. The person
using the biometric system wants the system to authenticate her and is
not likely to make faces, wear dark glasses, turn sideways, or otherwise
try to fool the system. And most important, the problem to be solved
is different. Instead of answering the question "Who is this random
person?" the system has to answer the much easier question: "Is this
person who she claims to be?"

We're far from the day where computers can reliably and independently identify people, but authentication is another matter. By the
way, one of the things the U.S. government didn't tell us about the
National Guard soldiers staffing the security checkpoints at airports
after 9/11 was that they all memorized a small list of faces that they
were watching for. That solution is considerably more effective than
having computers do it, but it's hardly a long-term solution.

. . . .

Authentication is a system, and making it work is much more than simply deciding how to authenticate someone. Following are some of the considerations necessary to create a secure authentication system.

First, where is the authentication stored? The people being authenticated have traditionally carried their own authentication information, but the person doing the authentication can also carry it. The latter method is more secure.

Let's say you present a photo ID to some guard, who compares how you look to the authentication information, your photo. This works, but it means that a forged ID—or a real ID with the original photo removed and a new one inserted—can fool the guard. However, if the guard has your authentication information on a computer, he can enter the serial number on your ID and not only look up your name, but also get a copy of your photograph. Then a forged ID is of no use. If the forged ID has the wrong name or the wrong picture, the guard should notice. (Many cruise lines use this system to identify passengers returning from shore excursions.) The system would be no less secure if you didn't have an ID card at all and just remembered your name or ID number. The guard would then type it in and get all the information he needs. The prevalence of easily available databases and good network connections makes this kind of authentication possible. In the U.S., all passport control desks have networked computers. In many jurisdictions, police cars are equipped with wireless computer terminals.

If ATM cards stored the customer's PIN on their magnetic stripe, an attacker who stole a wallet with an ATM card could attempt to recover the PIN from the card. Instead, the PIN is stored in the bank's database. The customer puts her card into the slot and enters her PIN, and the ATM sends the information to the bank's computer, which verifies the PIN. No secret information on the card, no risk of attack. If, however, attackers managed to get into the database, they would have PIN information for all of the bank's customers. The system is more secure against one kind of attack, and more brittle against another.

Another consideration is whether the authentication is being done remotely. Authentication across a network of some sort is more difficult than authenticating someone standing in front of you, simply because the authentication mechanisms are easier to fake and harder to verify.

Imagine a fanciful system that uses face recognition as a biometric. The system regulates access to a building. A guard sits in a room, unable to see the person who wants admittance. "In order to gain authorization, take a Polaroid picture of yourself and slide it through this small slot. We'll compare the picture with the one we have on file." This kind of system is easy to attack. To masquerade as someone, take a Polaroid picture of him when he's not looking. Then, at some later date, slip it through the slot instead of a photo of your face and fool the system. This attack works because while it is hard to make your face look like another person's, it's easy to get a picture of another person's face. And since the system does not verify that the picture is of your face (the guard cannot see you), only that it matches the picture on file, it can be fooled.

Similarly, an attacker can fool a signature biometric using a photocopier or a fax machine. It's hard to forge the vice president's signature on a letter giving an attacker a letter of credit, but it's easy to cut a vice president's signature out of another letter, paste it on the letter of credit, and fax it to the bank. It won't be able to tell that the signature was cut from another document, especially with modern digital editing software. Prisoners have managed to effect their own release by sending a fax of a forged document to prison officers.

For biometrics to work, the verifier must establish that the biometric matches the master biometric on file and that the biometric came from the person at the time of verification. If the system can't verify both, it is insecure. It's easy for a remote system to verify that the biometric belongs to the correct person, but it's much harder to verify that it wasn't taken from the person at some earlier time. For example, a security system that relies on voiceprints can be fooled by a tape-recorded voice.

Lots of authentication mechanisms fail this way, simply because they're not designed to be used remotely. More and more, I am asked to provide a photocopy of my identification—generally, a driver's license—when registering for something by mail. This kind of verification is a forger's delight, because he doesn't have to worry about embedded chips, microprinting, special inks, or intricate holograms that won't show up in a photocopy anyway. This is also an example of bad security reuse: The designers of the security countermeasure never intended it to be used in this way—by photocopy—and such a use loses most of the security of the original countermeasure.

Security issues surrounding token expiration also highlight the need for procedures for enrollment and revocation. In this chapter I have used the term "authorized circumstances" when explaining how security barriers are penetrated. All of the systems discussed presume that someone gets to decide who and what is authorized. Someone has to decide who gets the key to the door or knows the combination to the safe, who is given a driver's license or passport, and who is allowed on the airport tarmac. Someone has to designate who is and who is not a trusted person and delineate what the authorized circumstances are.

And in a very real sense, an authentication system is only as secure as the enrollment and revocation systems. We believe driver's licenses to be valid identification because we believe that people can get a driver's license only in their own name. This is an example of transitive trust: By trusting the license, we are trusting the people who issued the license and the procedures by which it was issued. If those people or procedures are subverted, it doesn't matter how secure the physical license is. Much of the debate about national ID cards ignores this important truism.

It is certainly possible to get an ID in a fake name, sometimes with insider help. And since it's a legitimate ID, antiforgery counter-measures are ineffective against this attack. Recently in Virginia, several DMV employees were issuing legitimate driver's licenses with fake names and selling them. (Two 9/11 terrorists were able to get Virginia driver's licenses even though they did not qualify for them.) Similar abuses have occurred in other states and with other ID cards. A lot of thought needs to go into designing the system that verifies someone's identity before a card is issued. Most important, a database may have to be interactive so that authorized persons may immediately alter entries to indicate that an ID holder's authorization status has changed—because of death or criminal activity, or even a change of residence. Because an estimated 5 percent of identity documents are reported lost or stolen each year, any database must be designed to reissue cards promptly and reconfirm the person's identity and continued qualification for the card.

Sometimes this notion of transitive trust is a way to sidestep liability. A bar might accept a driver's license as proof of age, not because it believes the license to be accurate, but because it knows that any misstatements on the license are not its fault. The agenda of the bar owner is not to ensure that no one underage drinks; the agenda of the bar owner is to follow the law.

I want to add a word about token expiration. Authorization tokens generally expire at the end of an authorization period. A monthly train pass, for example, expires at the end of a month. Some authorization tokens have no expiration date at all; a postage stamp is good forever (or, at least, until the government fails or inflation renders it worthless). Temporary badges that allow visitors into buildings might expire after a day, or a week. In my state (Minnesota), a driver's license expires after four years, and in order to get a new one, I have to take an eye test. My authorization to drive lasts a lifetime, as long as my vision stays okay and I don't get numerous DWI convictions. (In many European countries, driver's licenses never expire.) If humans were a species whose vision deteriorated much more rapidly, our driver's licenses might expire more frequently.

Authentication tokens don't have to expire until shortly before the authentication is likely to fail. Most passports around the world are good for ten years; that's about the right length of time until you need a new picture. Children's passports last only five years. That makes sense, because children's faces change faster. Signature cards on file with your bank never expire, although the bank will ask you to update your card if your signature changes substantially. Fingerprints, wherever they are on file, don't start failing until you get old; then they're practically useless.

Identification tokens never have to expire. There's no reason for a bank to change your account number. Nor is there a reason for the phone company to change your phone number. (Originally phone numbers were fixed to location, because they contained routing information. But that doesn't matter anymore.) This matter of expiration dates can be amusing when a token is used for a purpose other than what it was intended for. Driver's licenses are often used as general identification tokens: to prove age at a bar or to board an airplane. If someone were to show her expired driver's license to an airline gate attendant, he would most likely purposefully check the expiration date—which, of course, has no significance because for this use the license is just an authentication token. The picture is still of her. Perhaps it no longer authorizes her to drive, but it doesn't mean it's a forgery.

But forgeries are a reason why tokens expire. If someone is able to forge a passport, for example, he can impersonate someone else. If someone can forge a credit card, she can buy things in someone else's name. Expiration dates serve two purposes: They ensure that the bearer doesn't keep his authorization past a set time period, and they limit the usefulness of stolen cards and forgeries... assuming, of course, that the expiration date can't just be forged and that the verifier compares the card's expiration date with the date in the database (as credit card verification systems do).

     • • • •

All tokens, whether for identification, authentication, or authorization, must themselves be authenticated. When an electric company technician knocks on my door to gain entry, I need to authenticate whether he is a legitimate employee or a burglar trying to impersonate one. I do this by determining two things: if the photo on the badge is the person wearing it and if the badge itself is real. The second is forgery detection, the authentication of an object, and is much more difficult.

Object authentication is also as old as our species. We authenticate food by look and smell before eating. Currency is probably the object we authenticate the most throughout our day, although we don't think about it and don't do a very good job at it. Modern currencies use all sorts of technologies to prevent counterfeiting: special paper, watermarks, color-changing ink, precision alignment, microprinting, security threads, holograms. The security of these technologies depends on two things: how difficult they are to forge, and how good the people doing the authenticating are at recognizing forgeries.

Quite a bit of object authentication is done by machine. Money is authenticated in vending machines, slot machines, and banks' currency sorting machines. There are all sorts of money-authenticating technologies—some cheaper than others—and the sensible choices in a particular application depend on the threat analysis. A vending machine might have a simple authenticator that validates the weights of the coins going through the slots and measures a few aspects—color, printing—of bills. Who cares if someone steals a package of Fritos? A slot machine will have more technology to detect forgeries; in this case, someone can make real money with a successful fraud. I know of one coin authenticator that uses a laser to make detailed analyses of the coins coming through the slot.

While effective, none of these devices is perfect. In 1995, twenty-five people were arrested in Las Vegas and Atlantic City for passing $1 million in $50 and $100 bills, printed on a Tektronix Model 540 color laser printer. People could spot the fakes, but some slot machines couldn't. The police were tipped off when a prostitute complained about a customer giving her three counterfeit $100 bills. Another group of criminals printed U.S. currency, good enough to fool the slot machines, on their own printing press. The casino's finance department detected the counterfeits, and the police simply waited for the criminals to return to the same machines to continue their fraud.

If the difficulty of obtaining blanks is the only countermeasure, then the system is brittle: Someone with a blank medical prescription pad could create fake prescriptions. A month after photo driver's licenses were introduced in New South Wales, Australia, someone broke into the Motor Registry office and stole a machine that made the supposedly unforgeable licenses, along with thousands of blank licenses. In more resilient systems, the inability to get a copy of the blank document is just one of several security countermeasures. These days, some doctors send an e-mail to the pharmacy where you want to have your prescription filled. When you arrive with the handwritten prescription, the pharmacy can confirm its validity with the e-mail. E-mail can be forged, too, of course, so this added resilience isn't foolproof.

Many objects are authenticated without the benefit of special antiforgery technologies. Much of the caviar imported into the U.S. from Russia, for example, is "forged": black market, or not fresh, or not the quality labeled. If you're in the antiques business, you need to be able to tell the difference between an ancient Etruscan textile and a modern copy. Often the only countermeasure that works is knowledge and experience. If you know more than the forger, then you're likely to detect forgeries. If you know less than the forger, and if the forger is good at what he does, you're likely to have the ancient Etruscan textile pulled over your eyes.

Until recently, authenticating objects and authenticating data were largely the same process; you'd authenticate a piece of data by authenticating the physical representation of the data. Right now you're holding a copy of *Beyond Fear*, by Bruce Schneier. How do you know that this is the same book I wrote? How do you know that someone hasn't sneaked in and changed some sentences? Such an attack might not be a big deal with this book, but what if this were a book of drug dosages or the financials of a company?

Authenticating a book depends on the difficulty of producing a forged copy. The paper—the letterhead, perhaps the watermark—and the signature authenticate the information in a letter. The card, the ink, and the time clock authenticate the punch-in, punch-out data on a time card. The data on a cash register tape is authenticated because the paper itself is deemed not to be a forgery. That's the reason it's on a continuous roll in the register. Traditionally, this process has worked marginally well. While it's difficult for an average person to produce a forged copy of the book you're holding in your hands, Third World publishers have been doing that sort of thing for decades. Even before modern copying and printing technologies, Indian presses regularly republished books. (One presumes they didn't deliberately change things—typos were common—but they certainly could.) Similarly, a well-funded attacker could very well forge a watermarked piece of paper, a time card, or a roll of paper register tape. The CIA regularly forges well-worn authentic-looking passports from a variety of countries.

Still, an attack like forging a book doesn't happen often. When a doctor consults the *Physicians' Desk Reference* for a drug dosage, she trusts the data in the book because she believes that the book itself is not a forgery and has been checked for accidental errors. Certainly it's possible that a murderer might substitute a forged copy of the book with altered dosages, but it isn't a likely method of attack. While the threat is there, there isn't much risk. On the other hand, in 1614 a phony Don Quixote novel appeared, allegedly written by Cervantes. (Cervantes actually included some of the phony characters and events in his legitimate sequel.) And in 2002, three phony Harry Potter novels appeared in Chinese, at least one of them—*Harry Potter and Leopard-Walk-Up-to-Dragon*—allegedly written by J.K. Rowling.

Because of technology, authenticating data by authenticating its physical representation has become more difficult. Desktop publishing and high-quality printing have made it easier to forge letters, booklets, documents, and the like. In computers, data is separated from the physical instantiation of the data. If this book is ever sold as an electronic book, how will you know that someone hasn't modified paragraphs? If you look up a drug dosage or a company's annual report on a Web site, how do you know the information is correct? Even more important, how do you even know that the Web site is authentic? More and more, people are being fooled because of their inability to authenticate things in cyberspace. In 2002, criminals set up a fake Web site for the Bank of America and fooled customers into giving them personal information that they could then use to defraud the people. Scams like this, involving fake emails and Web sites, abound.

The only solution is to invoke trusted people. You have to trust that the e-book seller isn't going to sell you a forged book or an error-filled *Physicians' Desk Reference*. You have to trust that the legitimate publisher of Harry Potter books (Scholastic in the U.S., the People's

Literature Publishing House in China) isn't going to sell you one written by someone other than J.K. Rowling.

There is one possible type of data authentication employed by owners who want to be able to detect if someone steals, copies, or uses data without authorization. Map companies regularly add fake features to their maps: a small town, a hill. Phone books and mailing lists have fake listings. The idea is that if someone copies the data wholesale and sells it as their own, the original data owner can prove the theft in court. Basically, this is a type of watermark for data. This countermeasure depends on secrecy, of course; if attackers know which data is fake, then they can remove it. There are also data watermarks for electronic files like images, music, and movies. Their security depends on how easy it is to remove the watermark, and sometimes on the secrecy of the watermark's existence in the first place.

. . . .

Authentication systems suffer when they are rarely used and when people aren't trained to use them. For example, if someone approaches you and says he's from the FBI, or Scotland Yard, or the Ministry of Defense, how do you verify that he is who he says he is? Do you know what one of their ID cards looks like? Could you identify a forgery? I know I couldn't. And there's a power imbalance; many people are reluctant to question a police officer because he might take offense and retaliate. Some years ago, a CIA agent approached me and wanted to ask me some questions. (No, I didn't help him. Yes, the CIA is going to be unhappy when agents read this paragraph.) I told him that before I would even believe that he was from the CIA, I wanted to see him at the CIA headquarters at Langley walking out of the turnstiles. I figured that if he could do that, he was legit.

Imagine you're on an airplane, and Man A starts attacking a flight attendant. Man B jumps out of his seat, announces that he's a sky marshal, and that he's taking control of the flight and the attacker. (Presumably, the rest of the plane has subdued Man A by now.) Man C then stands up and says: "Don't believe Man B. He's not a sky marshal. He's one of Man A's cohorts. I'm really the sky marshal."

What do you do? You could ask Man B for his sky marshal identification card, but how do you know what an authentic one looks like? If sky marshals travel completely incognito, perhaps neither the pilots nor the flight attendants know what a sky marshal identification card looks like. It doesn't matter if the identification card is hard to forge if the person authenticating the credential doesn't have any idea what a real card looks like. Uniformed sky marshals would be much more secure against this kind of failure because the uniforms would be seen frequently. On the other hand, putting a sky marshal in uniform is like putting a huge bull's-eye on his chest. This is a classic security trade-off.

Perhaps the best solution is to require sky marshals to show their identification cards to the pilots and flight attendants whenever they board an airplane. Then, assuming the cards are hard to forge, this failure would not happen. If there were an identification dispute, the flight attendants could point to the authentic sky marshal. And since the flight attendants already have the trust of the passengers, they would have credibility. If the flight attendants are all incapacitated ... but I'm not going to go there. No system is ever foolproof.

Many authentication systems are even more informal. When someone knocks on your door wearing an electric company uniform, you assume she's there to read the meter. Similarly with deliverymen, service workers, and parking lot attendants. When I return my rental car, I don't think twice about giving the keys to someone wearing the correct color uniform. And how often do people inspect a police officer's badge? The potential for intimidation makes this security system even less effective.

Uniforms are easy to fake. In the wee hours of the morning on 18 March 1990, two men entered the Isabella Stuart Gardner Museum in Boston disguised as policemen. They duped the guards, tied them up, and proceeded to steal a dozen paintings by Rembrandt, Vermeer, Manet, and Degas, valued at $300 million. (Thirteen years later, the crime is still unsolved and the art is still missing.) During the Battle of the Bulge in World War II, groups of German commandos operated behind American lines. Dressed as American troops, they tried to deliver false orders to units in an effort to disrupt American plans. Hannibal used the same trick—to greater success—dressing up soldiers who were fluent in Latin in the uniforms of Roman officials and using them to open city gates.

Spies actually take advantage of this authentication problem when recruiting agents. They sometimes recruit a spy by pretending to be working for some third country. For example, a Russian agent working in the U.S. might not be able to convince an American to spy for Russia, but he can pretend to be working for France and might be able to convince the person to spy for that country. This is called "false flag recruitment." How's the recruit going to authenticate the nationality of the person he's spying for?

Authenticating foreign currency has a similar failure. About fifteen years ago, I was in Burma and a street-corner currency changer tried to change old, worthless Burmese money for my American dollars. The only reason I wasn't taken is that earlier my bus driver had warned me of the scam. Otherwise, how would I know what real Burmese *kyat* looked like? Some Las Vegas taxi drivers give casino chips from defunct casinos as change to unsuspecting passengers. Familiarity is resilient; novelty breeds brittle security.

In 1975, Stephen Holcomb walked into a Traverse City, Michigan, bank with a German 100,000-mark note, printed in 1923. The foreign exchange teller dutifully cashed the note, and Holcomb walked out with $39,700 cash for an otherwise worthless piece of paper. And in 2002, someone used a fake $200 bill with a picture of George W. Bush on the front to buy a $2 item at a Dairy Queen in Kentucky. The clerk accepted the bill and even gave him his $198 in change.

In Chicago in 1997, someone spent French franc traveler's checks as if they were dollars. The store clerks dutifully inspected the traveler's checks to make sure they were authentic, but didn't think to check the type of currency. Since the French franc was worth about 17 cents back then, the attacker made a tidy profit. Another scam centered around substituting French francs for Swiss francs. Still another involved writing a check with the correct numerical dollar amount and a smaller amount in longhand. The person receiving the check probably just looks at the number, but the bank pays what the words say. All but the Dairy Queen story are examples of I'm Sorry attacks; even Holcomb might have gotten away with saying that he didn't know German history.

If someone doesn't know what characteristic of the object to authenticate, that's the weak link in the system: Nothing else matters. I have a friend who has, on almost every one of the many flights he has taken since 9/11, presented his homemade "Martian League" photo ID at airport security checkpoints—an ID that explicitly states that he is a "diplomatic official of an alien government." In the few instances when someone notices that he is not showing an official ID, they simply ask for a valid driver's license and allow him to board without a second glance. When he noticed that the gate agents were scrutinizing expiration dates on IDs, he simply made another version of his Martian League card that included one.

Example: Face Scanning in Airports

The trade-offs for automatic face-scanning systems in airports are more compli-cated than their rates of active and passive failures. Let's go through the five steps.

Step 1: What assets are you trying to protect? We're trying to protect air travel-ers, and people in general, from terrorists.

Step 2: What are the risks to those assets? The risk is that a known terrorist will board an airplane. Even if he is boarding the plane without any ill intentions, he might have future terrorist plans, and we would like to identify, stop, and interrogate him.

Step 3: How well does the security solution mitigate those risks? Not well. There are about 600 million airplane passengers in the U.S. per year. Any system that has any hope of catching real terrorists will falsely accuse hundreds of thousands of inno-cent people per year. The system won't be trusted by the security screeners, and they'll most likely end up ignoring it.

Step 4: What other risks does the security solution cause? We have to secure the database of faces. If the database becomes public, then terrorists will know whether they're in the database and what their picture in it looks like, so they can modify their looks. Also, we need trusted people to manage the database. Anyone who is author-ized to modify the database could add people in order to harass them or could remove people. We need to secure the process that designs, develops, and installs the system. We need to secure the operational system so that it can't be disabled while in use.

Step 5: What trade-offs does the security solution require? Money, first of all. This is a very expensive system to install, and an even more expensive one to adminis-ter and maintain. People need to be hired to respond to all those false alarms. Then there is the inconvenience and the invasions of privacy for all of those innocents flagged by the system, and the cost of lawsuits from those harassed or detained. What happens if someone is wrongfully included in the database? What rights of appeal does he have? It's up to society to decide if all of that is too great a price to pay to fly in this dangerous age.

A system like this is clearly not worth it. It costs too much, is much too intrusive, and provides minimal security in return. In fact, if a system forces guards to respond to false alarms constantly, it probably reduces security by occupying them with use-less tasks and distracting their attention. All field trials for these kinds of systems have been failures, and the only major proponents of face-recognition systems are the companies that produce them. Their agenda is to convince everyone that the technol-ogy works, and to sell it.

Example: Biometric Access Control

The failure of automatic face scanning at airports doesn't mean that all biometrics are useless. Let's consider biometric access control—an example of biometrics being used as an authentication tool.

Step 1: What assets are you trying to protect? Whatever assets are behind the barrier being protected by this system.

Step 2: What are the risks to those assets? The risks are that unauthorized persons will get access to those assets.

Step 3: How well does the security solution mitigate those risks? Pretty well. A person identifies himself to an access-control system (generally with an access code), and we want the system to authenticate that he's really who he says he is. Some attacks can fool the systems—one Japanese researcher recently showed how to fool most commercial fingerprint readers with a fake gelatin finger, for example—but biometric technology is up to this task. Biometrics are convenient: no access codes to forget or keys to lose. The system needs to have good failure procedures for when someone legitimate is not recognized—perhaps she has a cut marring her fingerprint—but with that in place, it's a good solution.

Step 4: What other risks does the security solution cause? The system must be trusted. There must be a secure process to register people's biometrics when they are allowed access, to update their biometrics when their access fails for some reason, and to delete their biometrics when they're no longer allowed access. Other potential problems are that the biometric could be stolen or an attacker could create a false biometric to gain access to the system. Remember: Biometrics are unique identifiers but not secrets. Making this system as local as possible—not sending the biometric over the Internet, for example—will minimize this risk.

Step 5: What trade-offs does the security solution require? Money. Biometric authentication systems are considerably more expensive than "something he has" solutions. On the other hand, this solution is more secure than giving everyone a key (or even a code to a combination lock). And it's much cheaper than posting guards at all the doors.

In many situations, this trade-off is worth it. Biometrics would be an effective addition to airport security, for example. I can imagine airline and airport personnel such as pilots, flight attendants, ground crew, and maintenance workers using biometric readers to access restricted areas of the airport. They would swipe a card through a slot or enter a unique code into a terminal (identification), and then the biometric reader would authenticate them. That's two forms of authentication: the card or memorized code and the physical biometric.

The difference between a system like this and a system that tries to automatically spot terrorists in airports is the difference between identification and authentication. Face scanning in airports fails as an identification mechanism not because biometrics don't work, but because of the huge number of false positives. Biometrics are an effective authentication tool but not, at this point, an effective identification tool.

Example: ID Checks in Airports and Office Buildings

About a year after 9/11, I visited a New York company. Its offices were in a large Midtown Manhattan building, and it had instituted some new security measures. Before I was allowed into the elevator area, the guard asked to see my ID and had me sign in. While I waited, I watched building employees wave their badges past a sensor. Each time it beeped securely, and they walked on.

Since 9/11, ID checks seem to be everywhere: in office buildings, at hotels, at conferences. Examining the countermeasure with the five-step process will shed some light on its efficacy.

Step 1: What assets are you trying to protect? The building and the people in it, by preventing anonymous people from getting into the building.

Step 2: What are the risks to those assets? Many and varied. Basically, we are worried that people might do mischief.

Step 3: How well does the security solution mitigate those risks? Given how easy it is to get a fake ID—hint: get a fake from another state or country; the guard is less likely to know what a real one looks like—and how inattentive the average guard is, not very well. I believe that you can get past the average guard with a fake corporate ID printed on a home computer and laminated at a nearby Kinko's, with a fake name and fictitious company. Even if you assume the guards are more on the ball, most high school students can tell you where to get a fake ID good enough to fool a bartender. And what happens if a person does not have ID? Does the guard keep him out? Does someone inside the building vouch for him?

Step 4: What other security risks does the security solution cause? The security problem I would be primarily concerned about is the false sense of security this countermeasure brings. If the guards are supposedly stopping the villains in the lobby, then everyone in the building must be trustworthy, right? Even the guards may be more likely to ignore their own intuition and what's going on around them, because they're too busy checking IDs. Bad security can be worse than no security, because people expect it to be effective, and consequently they tend to let down their guard.

Step 5: What trade-offs does the solution require? Outfitting everyone working in the building with an ID card can't be cheap, and some office has to take on the job of issuing IDs to new employees, confiscating IDs from departing employees, and replacing lost cards—all activities fraught with their own potential security problems. Then there's the fact that electronic readers must be acquired and maintained. There's the inconvenience to visitors. And there's the continuing erosion of personal freedom.

By this analysis, the countermeasure is not worth it. Individual offices have always authenticated people to the extent they needed to, using receptionists, locks on office doors, alarm systems, and so on. An additional centralized ID-checking system is security theater at its finest.

ID checks at airports are no more effective, but I can explain why they exist: It's the agenda of the airlines. The real point of photo ID requirements is to prevent people from reselling nonrefundable tickets. Such tickets used to be advertised regularly in newspaper classifieds. An ad might read: "Round trip, Boston to Chicago, 11/22–11/30, female, $50." Since the airlines didn't check IDs and could observe gender, any female could buy the ticket and fly the route. Now that won't work. Under the guise of a step to help prevent terrorism, the airlines solved a business problem of their own and passed the blame for the solution on to FAA security requirements.

Example: National ID Cards

Currently the U.S. doesn't have a national ID card, but people are talking about it. Some want the federal government to issue these IDs; others want to standardize state driver's licenses into a de facto national ID card. About a hundred countries already have official, compulsory, national IDs that are used for a variety of purposes.

A national ID card system is actually very complicated and requires four major components:

- A physical card with information about the person: name, address, photograph, perhaps a thumbprint. The card might also include place of employment, birth date, religion (as Indonesia's has), names of children and spouse, even health-insurance coverage information. The information might be in text on the card or might be encoded on a magnetic strip, a bar code, or a chip. The card would also contain some sort of anti-counterfeiting measures: holograms, microprinting, special chips, and the like.
- A database somewhere of card numbers and identities. This database would be accessible, in authorized circumstances, by people needing to verify the card, just as a state's driver's license database is utilized today.
- A system for checking the card data against the database. This would be some sort of computer terminal that law enforcement would use. Presumably there would be a wireless version for patrol cars.
- Some sort of registration procedure to verify the identity of the applicant and the personal information, to add the information to the database, and to issue the card.

Step 1: What assets are you trying to protect? There's no specific answer. In general, we are trying to prevent crime and terrorism. So the assets are about as general as they can be.

Step 2: What are the risks to those assets? Crime, terror . . . the standard risks.

Step 3: How well does the security solution mitigate those risks? The "solution" is identification and authentication; national ID cards would allow someone to verify another person's identity. Doing this might improve security in certain isolated situations, but would have only a limited effect on crime. It certainly wouldn't have stopped the 9/11 terrorist attacks—all of the terrorists showed IDs, some real and some forged, to board their planes—nor would it have stopped the anthrax mailings of 2001 or the DC snipers of 2002.

In practice, IDs fail in all sorts of ways. The cards can be counterfeited. The manufacturers of these cards will claim that their anti-counterfeiting methods are perfect, but there hasn't been a card created yet that can't be forged. Passports, driver's licenses, and foreign national ID cards are routinely forged.

Human nature dictates that those verifying the card won't do a very good job. How often does a bartender—or an airport screener, for that matter—fail to look at the picture on an ID, or a shopkeeper not bother checking the signature on a credit card? How often does anybody verify a telephone number presented for a transaction?

And not everyone will have a card. Foreign visitors won't have one, for example. People will lose or forget their IDs regularly, kids especially. Except in isolated circumstances, systems can't be put in place that *require* the card.

The biggest risk of a national ID system is the database. Any national ID card assumes the existence of a national database, and that database can fail. Large databases of information always have errors and outdated information. Much of the utility of national ID cards assumes a pre-existing database of bad guys. We have no such database; all attempts are riddled with errors and filled with the names of innocent people. The verification mechanism can fail. It can be eavesdropped on. It can be used fraudulently. And the registration mechanism can fail; there always seems to be a way for someone to get a valid ID card through fraudulent means.

Given these failure modes, how well do IDs solve the problem? Not well. They're prone to errors and misuse, and worse yet, they are likely to be blindly trusted even when they're wrong.

Step 4: What other security problems does the security solution cause? The failures discussed above highlight many security problems, but the biggest is identity theft. If there is a single ID card that signifies identity, forging it will be all the more damaging. And there will be a great premium for stolen IDs (stolen U.S. passports are worth thousands of dollars in some Third World countries). Biometric information—whether it is photographs, fingerprints, or iris scans—does not prevent counterfeiting; it only prevents one person from using another person's card.

There is also the risk that the information in the database will be used for unanticipated, and possibly illegal, purposes. Finally, there isn't a government database that hasn't been misused by the very people entrusted with keeping that information safe—that is, the government itself. Police officers have looked up the criminal records of their neighbors. State employees have sold driving records to private investigators. Motor vehicle bureau employees have created nicely profitable side businesses selling fake driver's licenses. And this kind of abuse doesn't even necessarily require the malicious actions of an insider. Sometimes the communications mechanism between the user terminal—perhaps a radio in a police car or a card reader in a building—has been targeted, and personal information has been stolen that way.

Step 5: What trade-offs does the security solution require? First, there's the cost. For each of almost 300 million U.S. citizens, cards with a chip and some anti-counterfeiting features are likely to cost at least a dollar each, creating the database will cost a few times that, and registration will cost many times that. And even these costs will be dwarfed by the amount spent on ongoing maintenance. Add database terminals at every police station—presumably we're going to want them in police cars, too—and the financial costs easily balloon to many billions of dollars. As high as the financial costs are, the social costs are potentially worse. This kind of system has been extensively abused in many countries and would be abused in the U.S., as well.

This just isn't worth it. National IDs could be useful under some limited circumstances, but they don't represent a smart way to spend our security dollars. They are much more likely to be used properly by honest people than by criminals and terrorists, and the 9/11 terrorists have already proven that identification isn't a useful security countermeasure. Given that there really isn't a risk that needs mitigating, given that ID cards don't solve the problem of identification nearly as well as proponents claim they do, given that national ID cards cause all sorts of other security risks, and given that they're enormously expensive in terms of both money and personal freedom, this is not a solution we should pursue.

Chapter 14

All Countermeasures Have Some Value, But No Countermeasure Is Perfect

Neither attackers nor defenders can afford to rest: In the long run, they are forever spurring one another on. Attackers develop new ways to defeat countermeasures, which in turn leads defenders to develop better countermeasures, which in turn leads attackers to develop still better attacks, and so forth. While this results in a kind of progress, different defenses, at the right moment and in the right mix, can be extremely effective. Even things as simple as running away, or having a plan in the event of attack, have their place in a defender's arsenal.

By the time the Berlin Wall came down in 1990, it presented an impressive series of countermeasures. As you read through the list, notice the defense in depth, the combination of prevention, detection, and response, and the overall resilience of the security system. From east to west, there were . . .

- 302 watchtowers, with armed East German guards.
- An initial barrier, either a 12-foot-tall smooth concrete wall or a 10- to 13-foot-high wire-mesh fence. The wall was intermittently outfitted with some kind of electronic warning device.
- A field of steel stakes, planted in the ground, known as Stalin's Grass.
- Barbed wire entanglements.
- 20 bunkers scattered along the perimeter.
- Hundreds of leashed guard dogs running along a rail.
- A 20- to 50-foot strip of sand-covered earth, designed to reveal footprints of anyone crossing. This strip was also mined.
- A ditch, 10 to 16 feet deep, designed to stop any vehicles that got that far.
- A road with specially designed noiseless armed patrol vehicles.
- An electric fence, six and a half feet high, outfitted with either acoustic or optical alarms.
- A narrow band of barren land, called the Death Strip.

- A final barrier, a concrete wall between 11.5 and 13 feet high. An asbestos cement tube topped this wall, designed to prevent escapees from grabbing hold of the top. This is the wall that was graffiti-covered on the Western side, the one featured on the nightly news worldwide.

Despite this interwoven system of protection, detection, and response, thousands of people still managed to escape Eastern Europe each year. Some flew over the wall. Some tunneled under it. Some were guards, trusted to defend the wall. Many traveled to a different part of the border, where the barrier wasn't nearly as impressive. Companies even advertised extraction services. One of them charged about $12,000 to smuggle a person out of Eastern Europe and offered quantity discounts to families and a money-back guarantee if the extraction failed. (Hans Ulrich Lenzlinger, the company's director, was assassinated in his Zurich home in 1979.)

All security countermeasures have some value, in some circumstances, but none of them provides absolute security. Figuring out which ones work and which ones don't in any particular circumstance means understanding the risks and trade-offs. The five-step process we've been using throughout this book is designed to standardize the mental process for making those trade-offs. It's worth repeating: No security countermeasure is perfect, unlimited in its capabilities, completely and permanently impervious to attack. No such countermeasure exists, and, I would argue, no such countermeasure will ever exist.

Even so, we continue to design, implement, and refine security countermeasures, since we know security has to be an ongoing process. It's a process driven by new assets and new uses of those assets, new attacks and new ways to defeat countermeasures, and technological advances in both attacks and countermeasures.

. . . .

A rabbit's primary defense is *running away*. It's a fine defense, a useful countermeasure that almost everyone has used at one time or another. But it only works, of course, if you can outrun your attackers—by being faster, by being able to run longer, or by being clever about losing your pursuers. (Raccoons are particularly clever, doing things like running into and out of streams or climbing onto dangling tree branches.) Most herbivores tend to win on distance, which is why carnivores spend so much effort trying to sneak up close to their prey.

Hiding is another of the most obvious methods of defending against attack. Some animals hide against predators. Someone might have a safe in which to store valuables, but that safe will most likely also be hidden within the house. Armies use artificial smoke to hide troop movements. During World War II, many cities imposed blackouts to hide possible targets from bombers. Some homeowners hide an extra set of house keys somewhere outside the house—under the doormat is the canonical location, but under a rock in the garden is more secure—in the event of an emergency. And in wartime, many people buried their household valuables when fleeing an approaching army. Hiding can be either something done all the time—the safe example above—or something done in reaction to an attack. Running away and hiding are countermeasures that often are implemented together.

Hiding works best, of course, when there are a lot of places to hide. It's a great security countermeasure to hide needles in haystacks, because there's so much more hay than needle. Hiding a large airplane in a haystack is much less effective, although people in Czechoslovakia successfully hid a small steam locomotive in a haystack during World War II. Hiding somewhere in the entire U.S. is much easier than hiding somewhere in a neighborhood. Hiding is also improved when whatever is hiding looks like the things it's hiding in. A needle is shaped vaguely like a piece of hay; a button will stand out more. Both animals and militaries have developed camouflage techniques to make it easier to hide things.

Invisibility is a form of hiding. Many safes have plastic disks in their mechanism that set the combination. The disks are plastic rather than metal to make them invisible to X-ray probing. Stealth aircraft technology tries to do much the same thing: make the aircraft invisible—or, at least, unidentifiable—to radar and other detection mechanisms.

Hiding something in plain sight is a variant of the invisibility defense. It's not just in Poe's "The Purloined Letter" where it works; sometimes it's a real defense. The 9/11 hijackers made efforts to blend in with the local neighborhood where they were living; if they'd hidden in their basements and come out only at night, they would have aroused more suspicions.

Hiding fails when it doesn't take the attacker into account. A countermeasure that allows you to hide successfully from a mugger might not work against a police officer, and techniques that work against the police might not work against an organized crime syndicate. If the approaching army has night vision goggles, then the defenders can't use the dark of night alone to hide themselves. As an extreme case, think of the story of ostriches hiding themselves by burying their heads in the sand. (They don't, by the way.)

As with many of these countermeasures, attackers can use the technique of hiding as well. The apotheosis of this idea is the Trojan horse, inside which some Greeks hid so as to open the gates of Troy for the invading army. Pharaoh Thutmose III used the same trick to take the city of Jaffa, hiding troops in sacks of grain and bringing them into the city by ship. Muggers hide and wait for unsuspecting victims. Terrorists hide bombs in all sorts of objects: cars, garbage cans, and the like.

. . . .

Fighting back is a time-honored reaction to attack. Martial arts teach self-defense: how to fight back (or how to get free and run away). Police, army units, and any other groups with guns are likely to respond to gunfire by firing back. In the animal kingdom, teeth, claws, or a poison sting are often better defenses than a hard shell or fast legs. In Greek mythology, the primary defense of the Gorgons was to turn anyone who looked at them into stone. (It was, literally, a curse rather than a defense; the trade-offs were really high.)

In the post-9/11 era of airplanes as missiles, fighting back has been discussed a great deal. Some advocate armed sky marshals on all, or some, airplanes. Others advocate arming pilots. Many expect the U.S. Air Force to shoot down any terrorist-controlled airplanes.

Some of the most surprising examples of fighting back come from the plant kingdom. You wouldn't think stationary life-forms could do much, but lima bean plants have figured out how to call in air strikes against their attackers. When they are attacked by two-spotted spider mites, lima bean plants emit a chemical distress signal. The distress signal helps in three distinct ways. One: It gets other, nearby lima bean plants to start sending out the same distress signal, even if they're not being attacked yet. (Note the resilience of this system.) Two: It repels other two-spotted spider mites. And three: It attracts carnivorous mites to land on the lima bean plants and prey on the herbivorous two-spotted spider mites. The distress signal can't actually summon the predatory mites—mites ride around randomly on air currents and can't control their movements—but it induces predatory mites that happen to land on the lima beans to stick around and look for food instead of moving on.

To a great extent, the particulars of a situation determine how much fighting back a defender can do. There's a fine line between fighting back as a defense or as a counterattack, and different countries have different laws regarding what sorts of fighting back are legal and what sorts are illegal. For example, it may be legal for you to shoot (and kill) a burglar who has broken into your house. Many countries, and many (but not all) states in the U.S., see lethal force employed against home burglaries and some other crimes as a perfectly legitimate defense. But the laws of the UK are different, and so are those of many other countries; the argument used is that citizens must not defend themselves, because lethal defense is a monopoly right of the state.

In warfare, fighting back is expected. If you're part of an organized crime ring defending your turf against another organized crime ring, fighting back is expected. But in other situations, fighting back may or may not be legal. Please check with your local authorities before you booby-trap your home.

    . . . .

Diversions are primarily an attacker's tactic, one designed to make the security system's detection and response capabilities ineffective. The idea behind diversions is that if attackers can make defenders commit resources in the wrong place, then the attackers can better exploit the point they are really attacking.

Picture this diversion scenario: It's the 2002 World Cup finals, Brazil is playing Germany, and Brazil is ahead 1 to 0 at minute 79. Kleberson makes a run down the right side. Then, across the top of the penalty box, he sends a pass into the middle. Rivaldo, the obvious target for the pass, steps up to take the shot, drawing defenders toward him, but then at the last second he steps over the ball. The ball rolls to Ronaldo—unmarked and undefended—who shoots it into the lower-right corner of the net. Goal for Brazil!

Diversions can be used by the defense, too. At 3,106 carats, a little under a pound and a half, the Cullinan Diamond was the largest uncut diamond ever discovered. It was extracted from the earth at the Premier Mine, near Pretoria, South Africa, in 1905. Appreciating the literal enormity of the find, the Transvaal government bought the diamond as a gift for King Edward VII. Transporting the stone to England was a huge security problem, of course, and there was much debate on how best to do it. Detectives were sent from London to guard it on its journey. News leaked that a certain steamer was carrying it, and the presence of the detectives confirmed this. But the diamond on that steamer was a fake. Only a few people knew of the real plan; they packed the Cullinan in a small box, stuck a three-shilling stamp on it, and sent it to England anonymously by unregistered parcel post.

This is a favorite story of mine. Not only can we analyze the complex security system intended to transport the diamond from continent to continent—the huge number of trusted people involved, making secrecy impossible; the involved series of steps with their associated seams, giving almost any organized gang numerous opportunities to pull off a theft—but we can contrast it with the sheer beautiful simplicity of the actual transportation plan. Whoever came up with it was really thinking—and thinking originally, boldly, and audaciously.

This kind of counterintuitive security is common in the world of gemstones. On 47th Street in New York, in Antwerp, in London: People walk around all the time with millions of dollars' worth of gems in their pockets. The gemstone industry has formal guidelines: If the value of the package is under a specific amount, use the U.S. Mail. If it is over that amount but under another amount, use Federal Express. The Cullinan was again transported incognito; the British Royal Navy escorted an empty box across the North Sea to Amsterdam—where the diamond would be cut—while famed diamond cutter Abraham Asscher actually carried it in his pocket from London via train and night ferry to Amsterdam.

Why is it secure to mail gemstones without any other security measures? Why is it secure to mail classified military documents? It's secure—although brittle—because there's so much mail going through the system that no one will ever find your package; nothing in your package or envelope will make it stand out from the hundreds at the local post office, thousands in a central shipping facility, or millions handled daily by the postal service. In addition, if you simply mail something, very few people need to be involved in packaging and posting it, so very few people need to be trusted with the secret.

Diversions are more effective when combined with hiding. Government checks are regularly stolen out of mailboxes—in countries that still mail paper checks and still have unlocked mailboxes—if those checks are in easily identifiable envelopes. If I were going to mail diamonds, I wouldn't put a sticker on it saying: "Treat Securely. Diamonds." I would just insert it into the flow of over 500 million letters and packages mailed in the U.S. each day. The U.S. military's rules for mailing classified documents requires packages to be double-wrapped: the inner envelope with all sorts of classified warning labels, and a plain manila outer envelope.

Another hiding/diversion variant was used in the U.S. MX-missile land-based ICBM program. Treaties with the Soviet Union limited the number of nuclear missiles each side could have. The U.S. wanted to make it harder for the Soviets to destroy them all—I know, in retrospect this sounds like a bizarre scenario—so it devised a sort of shell game. The missiles were kept on special underground trains that traveled back and forth between different launch silos. There were many more silos than missiles. Since the Soviets would never know which silo held a missile at any time, they would have to target each silo independently and therefore waste many of their own warheads.

This is the sort of thinking behind decoys: three identical black limousines in a presidential motorcade, or Saddam Hussein's handful of body doubles who took his place at different public functions. During World War II, General Patton was given an entire fake army to pretend to invade Calais. Midas cichlid fish lose a lot of their young to predators and actually go out and kidnap the fry of other Midas cichlids, or even other species of fish, to use as decoys. Predators are just as happy to eat the adopted fry as the parents' own young, so as long as the larger school of fry doesn't attract more predators, more of the parents' own young will survive.

When squid are attacked they can produce pseudomorphs, blobs of ink held together with mucus. The idea is that the predator gets confused and pursues the pseudomorph while the real squid escapes. This tactic works because the squid has a complementary countermeasure: It can swim away at high speed. A flower can't use that trick to defend itself.

Similarly, submarines deploy drones to try to confuse torpedoes, and military aircraft have diversionary defenses against surface-to-air missiles: chaff to distract laser or radar guidance, and flares to confuse a heat seeker. You can play this game at home by installing a large wall safe for half your jewelry, then keeping the other (and perhaps the better) half in an old shoebox on the floor of a closet. It's diversion and compartmentalization combined in one security system.

• • • •

Deception is another important security countermeasure. Killdeer, a North American shore bird in the plover family, fake a broken wing to lead predators away from their nests, combining deception with diversion. While effective, the interaction of this security countermeasure with another fails badly. Killdeer lay their eggs on the ground among stones. The eggs are speckled and can be effectively invisible. The easiest way to find a nest is to find killdeers faking injury; they only do it near their nests. As with many countermeasures, this tactic works only against some attackers. Predators are fooled by the bird's deception, but against a human attacker who understands the bird's ploy, the security system fails.

When attacked by a predator, an opossum lies still, hardly breathes, and smells bad. The predator, thinking the opossum is dead, turns its attention elsewhere. When the predator leaves, the opossum gets up and walks away. Many of the opossum's predators don't eat already dead animals (as a preventive countermeasure against disease); they only eat animals they have killed themselves. Of course, the defense fails in the face of a carrion eater.

There are many more examples of deception in the natural world. Some insects have evolved to mimic other, poisonous, insects. Many flies have evolved to mimic bees, for example. Many species of fish have "false eyes": large spots on their side that look like the eyes of much larger fish. The mimic octopus can impersonate other creatures—lionfish, flounder, mantis shrimp, and sea snake—by changing its color, texture, arm position, and body language. Monarch butterflies taste nasty to predators, so several other species of butterfly mimic monarchs. King snakes are harmless, but have colorful bands, just like deadly coral snakes. Militaries sometimes do the same thing, having soldiers wear enemy uniforms to evade detection. The tradeoffs are significant, though; a captured soldier is a prisoner of war, but a captured soldier in an enemy uniform is shot as a spy.

Deception is a mainstay in the history of warfare. First-century A.D. Roman public official and soldier Julius Frontinus devoted a quarter of his book *Strategematon* to various ruses of war. Having a few horses drag branches behind them to create dust, so that a distant enemy thinks it is a large cavalry force, was a common deception technique for thousands of years. Similarly, armies would light many campfires or broadcast simulated radio traffic to inflate the assessment of the number of troops. During World War II, submarines on both sides tried the trick of releasing oil and debris when under attack, pretending that they were disabled underwater.

In the 1800s, rural banks sometimes installed a decoy safe in an obvious location, in an effort to fool bank robbers. The robbers would break into the safe and steal the small amount of money there, oblivious to the real safe in a more obscure location. One Internet security tool is a *honeypot*, a decoy computer network designed to be found first and look attractive, and thus lure hackers away from the real corporate network.

Some deceptions are much less elaborate. Many Americans traveling abroad pretend to be Canadians. With all the rampant anti-American sentiment in some countries, this is a perfectly reasonable security countermeasure. While it fails immediately when passport identification is required, it works for any casual interactions.

The effectiveness of any deception depends on two factors: the ability of those who know about the deception to keep quiet, and the preconceptions of the attacker. The first factor is obvious; deception requires secrecy. The second is subtler. The more the deception is in line with what attackers expect, the more they are likely to believe it. False eyes on small fish fool larger fish because they expect large eyes to belong to fish they should avoid. A second safe fools bank robbers only if its contents make it look like the real safe. If it were empty and unused, the robbers would most likely suspect there was another safe and start looking. Similarly, if you're being mugged, and the attacker finds only a few dollars and no credit cards in your wallet, he's going to suspect that you are wearing a money belt. And how successful you are at pretending to be from another country will depend on your complexion, facial features, and accent.

In 341 B.C., the kingdoms of Wai and Chao attacked the Han Kingdom in China. General T'ien Chi, of the Chi'i kingdom, marched to aid the Han with an army of a hundred thousand. Pretending to lose a significant percentage of his forces to privation and desertion, he had his armies light fewer and fewer campfires each night. The Wai and Chao commanders, who already had a low opinion of the Chi'i army, believed the ruse and were tricked into attacking prematurely. Chi ambushed the enemy and ended the war. Deception can be a powerful security countermeasure if used properly.

．．．．

Security seals are another kind of barrier. Think of a security seal on a bottle of medicine: The point of the barrier is not to prevent someone from opening the bottle, but to prevent someone from opening the bottle and then undetectably replacing the seal. There are all sorts of these seals: on food packaging, on the minibar in hotel rooms, on various government-calibrated measuring devices. The Federal Reserve wraps stacks of new bills so someone can't take one undetectably. Sometimes, instead of being called tamperproof seals, they're called tamper-evident seals: a better name.

Security seals, as a rule, are not really preventive countermeasures; instead, they're aids in an audit, since they come into play after the attack has already taken place. If someone breaks open a seal, the next person who comes along should know about it. Exactly how effective seals are depends on the threat model. Over-the-counter drugs have better seals than prescription drugs; since there were known instances of people poisoning bottles on store shelves, drug companies desperately need consumers to view these products as safe. Envelopes (the kind you mail a letter in) have a similar security model, but are secure only against a casual attacker—the kind who's a total amateur but can't control the urge to open your mail—think of, say, your kid brother. Against a professional—say, someone who works for a major governmental organization and routinely inspects other people's mail—an ordinary envelope is the proverbial piece of cake: She can open it easily and undetectably. (At the CIA, these specialists are known as *flaps and seals* experts.)

Food packaging falls somewhere in the middle. I remember being in a hotel once and, because there was simply no other option, deciding to eat a package of crackers from the hotel minibar. I was surprised to find that the seemingly sealed package had been surreptitiously opened from the bottom by someone who ate half the package and carefully replaced it on the shelf, with the opened seal on the bottom. The packaging seal failed as a detection countermeasure, because the person who refilled the minibar didn't bother checking it. (Some minibars have movement detection, so you can't drink the $4.00 can of Pepsi and later replace it with a can you bought in a store for 50 cents.)

These seals can also be found on various meters: postage meters, taxi meters, gas-pump meters, and so on. Old-style mechanical postage meters posed an interesting security problem: Anyone could tamper with the meter and get free postage. Traditionally, the procedure was that the customer brought the meter to the post office and paid for postage, and the postal clerk set the dials. He then sealed the meter with a tamper-evident wire, so that if the customer opened the meter she'd be detected the next time she brought the meter to the post office. Additional security came from an audit log book that the postal clerk wrote in, and the customer kept. This wasn't a great security system—if the customer never brought the meter in, she'd get free postage for life—but it wasn't an opportunity for high-value fraud. Newer-model meters are computerized and refilled remotely by telephone; the customer buys a code that is entered into the machine to refill the postage.

Seals are only as good as the verification mechanism. If someone doesn't audit them regularly, or accurately, they are ineffective as a security countermeasure. The original Tylenol poisoning case took place before tamper-evident packaging, but the attacker was so clumsy that his tampering was obvious nevertheless. The people who mistakenly took the poison didn't notice, probably because they weren't expecting this new kind of attack.

. . . .

If the asset you're protecting can be copied—think, for example, of music CDs, software, electronic documents, even just collections of data—then *replication* can be an effective security countermeasure. It doesn't protect the data from falling into the wrong hands, but it protects you from losing the data. (Again, it is important to understand exactly what assets, and what functionality of those assets, you are protecting.)

Biological systems latched on to this security countermeasure early in their evolution. Organisms put considerable effort into propagating their DNA. If each organism had just one offspring, odds that it would survive long enough to have its own offspring would be rather low. That's why many organisms have many offspring. A female lobster can lay 10,000 to 20,000 eggs at a time, but only ten to twenty survive four weeks after hatching. Nevertheless, the odds of any given lobster propagating its DNA are, on the whole, quite reasonable.

There are actually two different replication strategies in the natural world. The one described above is to produce many, many offspring and not spend much time nurturing any of them. The other strategy is exemplified by mammals: produce fewer offspring and spend more time nurturing them into adulthood. Both replication strategies work because biological systems are geared more to the survival of species than to the survival of individuals. Think of the large herds of gnu and other antelope that once wandered around Africa. If you are part of a herd of 50 million antelope, then the fact that there are 10,000 lions circling the edge of the herd doesn't matter all that much. The odds of any given gnu being eaten—at least before having offspring—are acceptably low, and the species will survive. (Actually, the size of the lion population is more critical to the survivability of the lions—too many lions compete, which is bad for every lion—which is why lions have fewer offspring than gnu.)

Computer backups also use the notion of replication. Smart users back up their data; if their computer is lost, stolen, or infected with a file-erasing virus, their data is safe. Most computer attackers don't intend to steal data; they're just wreaking mayhem or stealing computer hardware to resell. Sure, losing a computer is irritating and will cost money to replace, but losing data could be devastating; without backups, there's often no way to replace it.

Organizations that rely on data processing for their business, like airlines and credit card companies, often have complete backup data processing centers and well-practiced switchover procedures. That way, if they lose their primary center due either to attack or accident, business continues. Backups are a good resilient security countermeasure.

The downside of replication is that it is less secure against theft. Instead of just one copy of the data to secure, now you have at least two. And you have to secure the replication process. It's a classic design trade-off; your decision will depend on which attack you're more worried about and the relative security costs.

Replication can protect against other types of attack, but only if the security system is trying to protect one's access to something, rather than trying to prevent the attacker from getting access. For example, if I need to ensure that I have access to a car, a reasonable security countermeasure would be to have two cars. That way, if one car is stolen, the system does not fail. It's an expensive countermeasure, to be sure, and one that works only if I don't care about which particular car I drive, as long as one is available.

If an attacker discovers a class break—a way to steal all cars, or a way to disable all data centers—then replication doesn't work very well. Biological systems address this problem by making each animal slightly different. A virus that kills one animal is unlikely to kill the entire species; the survivors will live to propagate, and the species—now resistant to the virus—will live on. Technological systems are often homogeneous, which makes them much more brittle as a result.

. . . .

Security notices that warn of the existence of a defensive countermeasure act as their own countermeasure, in addition to any effects of the countermeasure itself. A burglar who sees a "Beware of Dog" sign might decide to rob another house and never actually encounter the dog. Guards, cameras, and alarms don't actually have to detect attacks in progress; their visible presence deters attacks before they happen. Security notices abound in the natural world: A rattlesnake's rattle, a coral snake's bright bands, a blue-ringed octopus's rings, and a bee's or wasp's stripes all signify that the animal is dangerous.

Animals tend to standardize on warnings. Instead of every kind of bee and wasp with a sting having a different warning coloration, most use black and yellow. The reason is that the only way predators learn to heed the warnings is to probe security and eat the animal. If different species standardize on the same warning signal, each species sacrifices fewer individuals to the predators' learning process.

This phenomenon has another emergent property: It's not necessary to have the actual countermeasure in place to get the deterrence benefits of the security notice. In Britain, for example, home alarm systems come with obvious red boxes on the outside of the house. One enterprising company makes and sells empty red boxes. People attach them to their house, and it looks as if they have an alarm system. Or they put a minatory "Beware of the Dog" sign on their front gate without actually owning a dog. A store owner I know installed a camera with a red blinking light over the cash register. He gets the deterrence effect of video surveillance without having to install a real system. In the security industry, these people are called *free riders*; they're trying to get the salutary effect of something without actually paying for it. In other circles, they might be considered clever improvisers.

But whatever you call it, the effectiveness of this kind of system depends on the quality of the fake. If burglars can tell that the red alarm box or blinking video camera isn't actually hooked up to anything, they're likely to be encouraged rather than deterred. What's more, the percentage of free riders also affects the efficacy of the system. If most alarm systems are real, then burglars are likely to steer clear of houses with alarm boxes. But if they realize that most alarms systems are false ones, they're more likely to start ignoring them or testing to make sure they're real.

It's also important to realize that the free rider phenomenon can spread a benefit without any action or even awareness on the part of the free riders. The LoJack automobile security system requires that a tracking device be secreted somewhere in a car. Cities where LoJack use is prevalent see a marked decline in car thefts across the board, because thieves begin to believe that everyone has one. After a certain percentage of people pay for the LoJack system, every other nonthieving car owner in the city receives the benefit from the system. It's important to consider that if everyone followed the optimal defense strategy—wait and let others buy it—then no one would buy it and no one would gain the benefit. The discussion of this phenomenon can veer deeply into sociological and economic theory, so we'll leave it here.

Several years ago, I was on an Air Force base at the same time Air Force One was parked there. Guards stood around the plane, and signs read, "Use of Deadly Force Authorized." When a *retaliation threat* acts as an effective preventive countermeasure, it's a form of deterrence.

This same idea—a public and explicit threat of retaliation—is behind the sign commonly found in stores: "Shoplifters will be prosecuted to the fullest extent of the law." Or the U.S.'s and USSR's clearly and repeatedly announced Cold War nuclear strategy: If one launched nuclear bombs against the other, the other would blow the first to smithereens. Or European hotels that demand to hold your passport during your stay, or libraries that demand to keep your ID card while you look at a particular reference book: "If you leave without paying, we will keep your valuable document, which—by the way—has your name on it." The library would rather have the book returned than keep your ID, and you would rather have your ID back than keep the book. In all cases, the implied retaliation is worse than the original attack, and the attacker is deterred.

Deterrence is also the idea behind positive bag matching, an airline security countermeasure. The idea is that bombers won't be able to check a piece of luggage and then not fly on the same flight, because the bag will be pulled off the plane if the passenger doesn't board. This will deter them, at least the nonsuicidal bomber. The airlines have gotten this right; it doesn't really matter if the bag actually flies on the same flight, as long as the passenger believes that it will. It's okay for the airline to route the bag on a different flight from the passenger, but it's not okay for the passenger to decide to fly on a different flight from his baggage.

. . . .

Being distasteful is an occasionally appropriate defense: making the target of attack—whatever it is—less palatable to the attacker. In A.D. 828, a group of Venetian merchants stole the body of Saint Mark from his tomb in Alexandria. To sneak the body past the Muslim customs inspectors at the harbor, the thieves (they're the defenders at this point) covered the body with pork. There's actually a mosaic on the outside walls of the Basilica de San Marco in Venice that shows the Muslim customs inspectors recoiling from the block of pork that hides the saint's body.

There are stories of Roman nuns who disfigured their own faces so that barbarian invaders would find them too ugly to rape. Cooking sherry is commonly salted, today to get around liquor licensing laws and taxes, but originally as a defensive measure against the servants drinking it. Milkweeds have a bitter alkaloid in their sap that most animals will avoid. Modern drug smugglers might learn from these stories and hide their goods in a shipment of dung, making it less likely that the customs inspectors search the shipment. The *Guide to Disneyworld* suggests that your baby stroller is less likely to be stolen if you tie a soiled diaper in a sealed plastic bag to the top. And for gross-out effect, I can top that. In high school, I knew someone who once ended an argument over whether he should share his potato chips by licking one and putting it back in the bag. Not very charming, but effective.

Chili peppers contain a chemical called capsaicin that makes them taste hot. Because of that, mammals—which would chew up the pepper seeds so they couldn't germinate—won't eat them. Birds, which pass the seeds whole through their digestive systems, don't find capsaicin distasteful.

A related defense against theft is to make the thing less useful; this defense is sometimes called *benefit denial*. You can buy a car stereo that comes in two parts: the stereo attached to the car and a faceplate with the controls that you are supposed to remove and take with you when you leave the car. Without the faceplate, the stereo is useless, and presumably there is less incentive to steal it. A related anti-shoplifting countermeasure is used for garments: tags attached to the garment that spread colored dye if removed improperly. Some bicycles have quick-release seats to facilitate removal, making the rest of the bicycle harder to pedal away. In 1991, the retreating Iraqis set fire to the Kuwaiti oil fields, a similar countermeasure.

This type of countermeasure pops up here and there. In New York City subways, the light bulbs have left-hand threads and do not fit in standard sockets. A year ago I stayed in a small hotel in Europe where the television needed a smart card in order to work. This security countermeasure both allowed the hotel to charge extra for television access and made the units much less attractive to thieves. And on the flight home, my airline provided me with a portable DVD player; all the DVDs came with the notice: "Disc will not play on commercial DVD players."

Along the same lines, an issue of *Mad Magazine* from the 1970s contained a sign for the outside door of your house: "Warning. These premises are protected by extreme poverty. There's nothing worth stealing." A good joke, and no different from a store sign: "No money kept in store overnight." Or, "Delivery person has less than $20." Note that these countermeasures not only protect the money, but also the clerk.

Some safes have a slot so employees can drop money in but cannot take money out. This countermeasure does several things. One, it provides security by limiting trust: Only certain employees can open the safe. Two, it makes the money secure against all attackers who don't know how to break open safes. And three, a sign that says "Clerk cannot open safe" acts as a deterrent and protects the clerk from coercion.

The city of Tokyo is designed to be hard to navigate. Most of the streets are unnamed, and the addresses are an impenetrable system of district (*ku*), zone (*chome*), block, and house, with the *chomes*, blocks, and houses not being numbered consecutively. (The houses are numbered by order of construction, more or less, and I never have learned the system of block numbering.) This system was ordered by Edo shogun Tokugawa, even though it was wildly inconvenient, because it created a defensive maze around the Imperial Palace. The countermeasure worked magnificently; nobody ever overran Tokyo. (On the other hand, the system regularly has active failures: Natives get lost all the time, too. Tokyo business cards have maps on their backs, and police officers constantly direct lost people.) Along similar lines, West Germany had plans to remove all street and road signs in the event of a Soviet Bloc invasion. When the city of London began putting up house numbers and street signs in the 1760s, people rioted because they didn't want strangers to be able to navigate through their neighborhoods.

Many of these countermeasures—like the dye tag on garments—depend on the secrecy of the tag-removal system to be secure. If shoplifters can get their hands on a portable tag-removal machine, as some have, the countermeasure no longer works. And there are limits: People who remove their car stereo faceplate when they're not driving may still get their windows smashed by thieves who find it easier to break the window than to check first.

A related countermeasure is *not being a target*. Some targets are more attractive because of who they are or what they do. Lone individuals walking down deserted and dark alleys are more attractive to muggers. Tobacco companies, oil companies, bioengineering companies, and nuclear power plants all attract attackers because of the businesses they're in. Other companies attract attackers because of shady corporate practices, perceived exploitation of people or resources, or excessive greed and incompetence. Some countries attract attackers because of their foreign policies, others because of their domestic policies. And certainly, the rich hegemony will always be unpopular.

One way to defend against attack is to not be someone who attracts attackers. For people, this might mean not flashing wads of money or expensive jewelry around when they're in a strange city; looking like a tourist is a security risk. (It's written in the *Tao Te Ching*: "... by not displaying what is desirable, you will cause the people's hearts to remain undisturbed.") For companies, this might mean spending money on public relations and generally being good corporate citizens. For nations, this might mean being responsible players on the world stage. On the other hand, this also means not taking unpopular stands on some issues.

This kind of thing happens in the natural world, as well. Adélie penguins have a problem eating: Both the things they like to eat and the things that like to eat them are found in the ocean. So they all bunch up on the shore, none willing to jump in, each nudging and jostling each other, until one jumps in or is pushed. Then they all rush to jump in at the same time. Ravens have a similar protocol for approaching dead animals that sometimes are not dead. After one brave raven takes a bite and doesn't get attacked, the others join in. I've seen people doing the same thing at a potluck buffet, nervously eyeing an unusual dish until some person is brave enough to try it.

Immediately after 9/11, some people said: "They must hate the U.S. for a reason. Let's try to learn what that reason is," or suggested that American arrogance and heedlessness in its foreign policy contributed to a political climate that nourished the seeds of terrorism. In the U.S., such talk was generally not well received. In Europe, that line of inquiry resonated more. Discussing this issue doesn't excuse the terrorists or somehow make their cause moral. It doesn't imply that the U.S. is somehow culpable for the attacks, or justify the terrorist's actions, or make them less abhorrent. A woman who avoids dark alleys because she's afraid of attackers isn't to blame if she ignores her own advice and the worst happens. Even so, she would have been safer had she heeded her own advice. And if the community adds more streetlights, it is not an admission of guilt on anyone's part.

Understanding the motivations of attackers and determining what you can do to make yourself a less attractive target is a reasonable line of inquiry. Like any other countermeasure, these can be deployed or not, depending on your assessment of the situation. Salman Rushdie would have been safer had he not written *The Satanic Verses*. That by no means implies that he should not have written it.

....

A security *protocol* is a series of steps that some trusted person carries out, steps designed to enforce some sort of security rules. They're everywhere in our daily lives.

There are protocols, for example, that merchants use before accepting a check. They may attempt to authenticate the customer's name and record that authentication information on the check. This information is recorded for forensic purposes, so they can better track the customer down if the check is bad. Some merchants won't accept low-numbered checks, on the theory that more bounced checks come from new accounts. A bank, on the other hand, has completely different protocols for determining if a check is good.

If a check bounces, merchants might decide to sue the customer. There is another protocol for this, one that involves judges and the court system. To avoid this, merchants might require a certified check. The bank would go through yet another protocol to certify the check. This process gives merchants some assurance that the check is good.

Our daily life is rife with security protocols. Many of them involve trusted people not directly involved in the interaction: The bank is the trusted party between the merchant and the customer, for example. These are known as *trusted third parties*, and they're a common security countermeasure in everyday commercial systems. Often they transfer risk from one party to another. In the example, the bank is acting as a trusted third party in the transaction. Merchants trust the certification on the check and trust that the bank will honor the check for its full amount. The customer trusts that the bank will keep the money for the check on hand, and not spend it on risky loans to Third World countries or telecommunications start-ups. Because they both trust the bank, merchant and customer can complete their transaction without trusting each other.

This system works—not because the bank is a solid institution backed by impressive-looking buildings and a spiffy advertising campaign, but because the bank has no interest in the individual transaction and has a reputation of trustworthiness to uphold. It will follow its protocol for a certified check no matter what. If the customer has enough money in an account, the bank will issue the check. If the merchant presents the check for payment, along with suitable identification and authentication, the bank will pay.

Lawyers often act as trusted third parties. They're executors for wills, mediators in contract negotiations, and filers of papers. In all of these cases, they act as a trusted go-between for parties who don't trust each other. Judges can act as trusted third parties, too. Their role is a little different. The bank is a trusted third party involved every time the customer and merchant exchange a check for goods. The exchange can't happen without the bank's involvement. Judges only get involved in the event of a dispute. If the merchant and customer are happy, the judge never hears about it.

Some protocols automatically enforce fairness. Coin flipping is a simple example of a protocol to pick one of two choices. There's also the cut-and-choose system, commonly used by people to divide the remaining pie in the refrigerator or a jointly purchased bag of marijuana. The idea is that one person divides the goods into two piles, and the other picks a pile for himself. What's interesting is that neither party can cheat. It is in the splitter's best interest to make two equal piles, because she knows that the other person gets to choose. And the chooser knows that he gets to choose, so he doesn't care how the splitter makes the piles. (For the more mathematically inclined, there are some very elegant variants for more than two people.) Auction protocols work along similar lines; they're designed to determine a fair price for something. And in every gambling game, one person shuffles but another cuts the cards.

A variant of this protocol can be used to set the value of real estate and other nonliquid assets. Determining the value of assets for tax purposes is a complicated and abuse-prone practice. Asset owners want the value set as low as possible. Governments have a complicated set of goals, depending on how corrupt or fair they are. There's a self-enforcing protocol that solves this problem nicely. Asset owners get to set the value of the asset at whatever they like. The government can then choose to either accept the tax calculated on that value or buy the asset at that price. Like magic, asset owners suddenly want very much for the asset to be valued fairly.

While this system won't work in practice for other reasons, it's a good illustration of how protocols can help secure systems: They simply remove potentials for abuse. Asset owners can't cheat the system, because the system itself prevents it. It also suggests that setting up protocols can create new risks, and that protocols should be analyzed as carefully as every other facet of a security system.

Protocols can make mediocre security countermeasures effective. A velvet rope in a nightclub—or yellow police tape at a crime scene, or a necktie hung on a doorknob—works as a secure barrier because people know not to cross it. Labeling food in a college dorm refrigerator is effective because the protocol is that people don't eat food labeled for someone else. In Inca cities, a horizontal pole across a doorway meant an uncrossable barrier; because people followed the rules, it was as good as a locked door. Of course, many attackers and attacks bypass protocols like this.

Protocols are also what turn good security countermeasures into something to laugh at. A perfectly sensible security protocol like "security screeners should confiscate all fake guns in airports" results in a security guard confiscating a G.I. Joe toy gun even though it's only an inch long. When protocols turn people into mindless drones, they are less effective.

Like protocols, *procedures* are steps that a trusted person carries out. But in security lingo, procedures are exceptions; they're the things that people do when a security event occurs. The police have procedures for negotiating with someone who has taken hostages, for approaching a stopped car, or for surveying a crime scene. A bank has procedures to follow when it suspects a customer of fraud, or if the vault door is discovered open in the morning. Protocols are the routines trusted people follow day to day; procedures are what they do in response to an anomaly.

All detection and response that involves people requires procedures. Response is difficult and complicated, and the times that a response is called for are likely to be stressful. If you want any hope of getting any of this right, you need to have procedures in place beforehand. For example, every guard has a series of security procedures that he is supposed to execute in various situations. The military has all sorts of procedures. On a more personal level, an obvious security procedure for a family is to have an agreed-on rendezvous point should the kids get separated while at a mall or amusement park. Procedures can be brittle and, like protocols, can turn people into mindless drones. But good procedures are essential to good security.

Planning is one of the most important aspects of security defense, especially for detection and response. Protocols and procedures need to be thought out, and are subtle and hard to get right. And the only way to get them right is to plan carefully. Around 12 percent of the 9/11 World Trade Center victims were firemen (343 of about 2,800), partly because their planning did not include the possibility of a catastrophic building collapse and partly because of poor communications procedures, especially among the off-duty firemen at the scene. Not that this was unreasonable; the situation was unprecedented. But certainly fire officials are planning for this possibility now. On the other hand, many of the companies whose offices were in the World Trade Center had comprehensive disaster recovery plans. For example, eSpeed—the subsidiary of Cantor Fitzgerald that runs electronic marketplaces—lost 180 employees and its main data center, yet was up and running again in two days. And because of the 1993 World Trade Center bombing, many companies had evacuation plans in place and people got out of the buildings quickly.

．．．．

The efficacy of response improves significantly with planning. Planned reaction is more effective. Mitigation and recovery are pretty much impossible without procedures in place. (Of course, not everything can realistically be planned for, but the more planning you do, the more resilience you've built into your security system.)

Vegetius, a fourth-century Roman military strategist, wrote, "Let him who desires peace prepare for war." This sentiment is true not only for military defense, but for every other kind of defense. You might not get it right if you plan, but anyone who doesn't plan can get it right only by chance.

You also have to make sure everyone knows the plan. The mitrailleuse was an early type of machine gun (crank-operated), and the secret weapon of the French army in the late 1860s. When it worked, it worked exceptionally well—mowing 'em down in windrows, as it were. But because it was such a secret, the French never deployed it during maneuvers so they could study its tactics. And they strictly limited the number of people who knew how to operate it. Come war with Prussia, two problems quickly developed: (1) no one knew the optimal tactics for the weapon in open combat, and (2) once the one soldier who was trained in the mitrailleuse was incapacitated, the weapon was effectively useless.

The mitrailleuse story illustrates benefits of *training and practice*. The more people who are trained, the more likely it is that someone will be able to do whatever needs to be done. And during practice, you can develop new and better procedures. Procedures aren't very useful without training and constant practice. When security events are rare, people don't expect them and often don't even notice them. Training is a way to deal with that aspect of human nature. When security events are rare, people don't gain experience dealing with them. Any system that is rarely used will fail when used. Practice solves that problem. One goal of training and practice is for people to react reflexively and instinctively in a crisis situation instead of having to ponder what to do next.

A new device used incorrectly can cause more damage than not having the device in the first place. Give someone a handgun without any training, and she's much more likely to hurt herself or someone else by accident than she is to defend herself successfully. Flying a plane is easy; it's the stalls, weather emergencies, getting lost, instrument failures, and those two essential but special circumstances—takeoffs and landings—that take all the training time.

In the mid-1990s, General Motors found that it was spending $3 billion annually on warranty repairs. About a third of that sum was wasted: Either the part didn't need repairing, or the repair didn't fix the problem. To solve this, GM created a training program to improve the quality of both diagnosis and repairs. Four years later, half of the company's repair force had gone through the training program. However, this produced no difference in diagnoses or repairs between the trained people and those who had not yet received the training. Cars are so complex and reliable that an average of seven months passed between technicians seeing the same failure twice. Skill decay set in, and they couldn't remember what they did to fix the problem last time.

Knowledge, experience, and familiarity all matter. When a security event occurs, it is important that those who have to respond to the attack know what they have to do because they've done it again and again, not because they read it in a manual five years ago.

. . . .

Finally there is *testing*, one of the hardest things in security. On one hand, it's impossible to test security. On the other, it is essential to test security.

As a system becomes more complex, with more variables and options and permutations, it rapidly becomes impossible to test. You can't think up all possible tests, and you can't afford to run all the ones you do think up. And if you can't test your security, you're forced to fake it. Perhaps you simplify the system and test it that way and then hope that your simplification didn't affect the security properties. Perhaps you can test what you think of and hope you haven't forgotten anything important. Whatever you do, you're going to miss something.

Imagine that you've decided that a 20-percent power surplus is necessary for security; the redundancy means that you're less vulnerable to an attack against any one power plant. How do you test that this countermeasure is effective? You can simulate an attack, but you don't know if it is a realistic simulation. And some attacks are simply too dangerous to simulate. You may be able to test some pieces of this system, but only a few. And even if you could, by some Herculean effort, test them all, you still would only be testing the pieces you knew about, against the attacks you could think of.

Some tests are impossible simply because of their consequences. You can't melt down a nuclear reactor to test the emergency cooling system. Instead you have to create the best simulation you can and hope that the simulation matches reality. Testing works best when you're worried about specific attacks and copycat attackers, the security system being tested is simple, and you want to test specific pieces of a system, not overall security itself.

When it comes to trusted people, though, testing is essential—not so much to see how good the security is as to see if procedures and protocols work, and if all the planning, training, and practice has paid off. If a security procedure depends on a trusted person calling someone else, for example, it makes sense to test that procedure periodically to make sure the phone works. There's no guarantee that the phone will work under attack, but it's more likely to if it's tested regularly. If the telephone line is cut regularly, perhaps by a squirrel, testing will uncover the problem. Regular operational testing makes readiness more likely. Perhaps the person plans to use a cell phone, and random testing demonstrates that his phone's battery is frequently very low, or the transmissions break up too much at that particular site, and communications might fail in the emergency. Testing will uncover this problem. Then whoever is in charge of security can create a procedure to prevent or correct that situation. Testing uncovers active failures much more easily than it uncovers passive failures.

In 1998, the Denny's chain closed its restaurants for Christmas for the first time since it was founded 35 years earlier. The catch was that because the restaurants were always open 24 hours a day, seven days a week, many of them didn't have door locks. This is the kind of problem testing will uncover.

Whenever security involves people, testing will provide considerable information. It might not tell you whether the security will work, but it will certainly give you insights into how it might fail.

. . . .

Execution is where the rubber meets the road. All the procedures and training in the world won't help if the plan isn't executed properly. The guards have to respond to the attack. The auditors have to notice the fraud. The army has to cut off the supply lines of the advancing army. Security is all about people—trusted people—and those people are expected to execute whatever security plan is in place to thwart the attackers.

After World War I, the French built the Maginot Line, a series of trenches and tunnels designed to be the ultimate defense against a World War I–style attack. Many people see the Maginot Line's failure during World War II as a failure of static security: The French were locked into a trench-warfare mind-set and did not respond to the technological advances of the previous two decades. This actually isn't correct. In World War II, the Maginot Line worked as intended. Its purpose was to channel any German invasion through Belgium, thereby ensuring that the fighting would not occur on French soil and guaranteeing that Britain would align itself with France against Germany. What happened in 1940 was that the Ardennes defensive arrangements that were planned were not implemented (i.e., they were supposed to send troops into the woods, to block access from that direction), leaving a defensive gap. So the failure of the Maginot Line was less a failure of vision, or even planning, than of execution.

On a smaller scale, the nine miners trapped in a collapsed Pennsylvania mine in July 2002 survived because they properly executed disaster procedures—stay together, seek high ground, and do not attempt to escape—so the rescuers could figure out where they must be and were able to sink a shaft to supply the warm compressed air that kept the miners alive.

Planning and procedures are essential. So are training, practice, and testing. But if you don't actually execute whatever you're supposed to do, you might as well not bother.

Example: Military Actions Against Terrorism

The U.S. military has been a powerful force in the fight against terrorism. The U.S. invaded Afghanistan (2001) and Iraq (2003), and sent troops to help combat terrorism in the Philippines, Indonesia, and Uzbekistan. On a smaller scale, in 1998 the U.S. launched a cruise missile attack against a suspected chemical weapons plant in Sudan (turns out it wasn't) and terrorist training camps in Afghanistan (turns out they were abandoned). On an even smaller scale, in 2002 we assassinated a terrorist leader in Yemen using a remotely launched Hellfire missile.

Is this a smart security policy? Does it increase security? We can't really analyze the specifics of most of these operations because we're missing many of the facts. But we can discuss what principles nations should follow as they try to combat terrorism around the globe.

(Note that this is a discussion of military action, not of the broader "war on terror." War on terror isn't a war in the normal sense of the word: one nation-state fighting another. The war on terror is a war against an idea, a tactic. It's a rhetorical war, and the phrase is more marketing than anything else.)

Step 1: What assets are you trying to protect? Potentially, everything.

Step 2: What are the risks to those assets? Terrorism. Specifically, those threats that originate from identifiable sources outside the countries where normal legal processes can be used to neutralize them.

Step 3: How well does the security solution mitigate those risks? The effects seem to be mixed. Military actions have resulted in the overthrow of the governments of Afghanistan and Iraq and the capture of several high-profile terrorist leaders, but it's hard to tell how much of the success is based on military actions and how much is based on our diplomatic and investigative actions (freezing Al Qaeda assets, for example). I don't have enough information to make an accurate assessment as to the efficacy of these actions, but I hope that someone in the U.S. military does.

Step 4: What other risks does the security solution cause? Anti-American sentiment is the serious one. And it's a really serious one. The more the U.S. violates the autonomy of other countries, accidentally kills innocent civilians, and uses military might to enforce its nonsecurity agendas, the more anti-American sentiment those actions cause and the more fertile breeding grounds become for anti-American terrorists. Increasing the prevalence (and zeal) of anti-Americanism increases the number of potential terrorists. Blowback—unanticipated negative consequences of an operation—is the major security risk in any antiterrorist military action.

Another risk is putting U.S. soldiers in places like Kabul and Baghdad and Lebanon where they're more easily reached by terrorists. And a third risk is getting dragged into more wars, whether they make any sense for security or not.

Step 5: What trade-offs does the security solution require? There are the basic ones you'll find in any military action: cost, loss of life (on both sides, military and civilian), economic volatility. And there's the basic ethical trade-off of any aggressive action: Is it the sort of thing that we, as a nation, want to do in the world?

What have we learned? The fundamental trade-off is between Steps 3 and 4: If a particular action fuels more anti-American hatred than it eliminates, then it's not worth it. If it deters people from becoming terrorists in the first place—this is similar to the rationale for spending so much on currency anti-counterfeiting technologies—then it is worth it. If sending troops into this or that country to kill some Muslim extremists fans the flames of anti-Americanism throughout the Muslim world and creates even more terrorists in the process, then it's not worth it. Inciting hatred is bad; being a good world citizen is good.

Of course, the decision to use military force is a national policy decision and not merely a security trade-off, but this analysis implies two concrete policy recommendations. The first is that unilateralism is a security risk. Using multinational organizations like the United Nations and international coalitions to wage war is very much in the U.S.'s self-interest, because it spreads the blame around. Liberating Kuwait with an international coalition is far more effective than invading Iraq with much of the world opposed; from the other side, a "coalition of the willing" looks an awful lot like a posse. Trying terrorists in a global venue like the International Criminal Court instead of in a U.S. court is in the U.S.'s self-interest, because it reduces the perception that American imperialism is at work and increases the perception that the world's nations are uniting against terrorism. Why give the various anti-American propaganda machines more fuel?

The second recommendation is that overt force should be used sparingly. Parading troops into a country is less desirable than sneaking a commando team in and out clandestinely, if both options can satisfy the same goal and if we decide that the goal is a good one.

The point here is not that military actions against terrorists are never justified, or even that the U.S. should not on occasion engage in unilateral military actions. The point is that both public and governmental policy debates on military action should include some form of this five-step analysis. These issues are bigger than security. For example, the decision whether to invade Iraq was a complicated one, with strong moral and political arguments on both sides that were not directly related to security. Military action needs to consider both arguments—not only within the context of the group the action is taken against, but also within the larger system of world opinion and future safety.

Chapter 15

Fighting Terrorism

The five-step process also applies to terrorism and counterterrorism. Because terrorism is largely a crime against the mind, the assets to be protected are hard to define. Terrorism is rare; while the threats are serious, the risks are lower than most people think; and responses are out of proportion to the actual risk. Most countermeasures are less effective than they appear, and many cause additional security risks. In an atmosphere of fear, keeping trade-offs commensurate with real risks is not easy.

As a society, international terrorism is probably the greatest security challenge we face today. It's a difficult topic because it's so emotionally charged, but the five-step process can help us approach it sensibly. Even if we don't have a specific countermeasure in mind, the questions highlight some of the particular problems we face defending ourselves and our countries against terrorist threats.

Recall the definition from Chapter 5: "A terrorist is someone who employs physical or psychological violence against noncombatants in an attempt to coerce, control, or simply change a political situation by causing terror in the general populace." Not everything being called terrorism in the media is terrorism, and it's important to use a precise definition.

Step 1: What assets are you trying to protect? "Everything and anything" is pretty close to the correct answer, but we can organize the assets in the following way.

- *People:* Any gathering of people is a potential terrorist target. Suicide bombers have blown themselves up in hotels, restaurants, college cafeterias, discos, buses, markets, and busy intersections. Terrorism has been directed against people on airplanes, but also against crowds waiting in airports. We have to worry about terrorism in concert halls, cruise ships, office buildings, commuter trains, shopping malls, and so on.
- *Public symbols:* Assets that have particular symbolic value are their own targets. The World Trade Center and the Pentagon were themselves targets beyond the people in them. Other potential

targets in this category include national monuments, government buildings, sports stadiums, bridges and tunnels, military installations, as well as events that draw large crowds of people, such as the New Year's Eve celebration in Times Square or the Academy Awards ceremonies. Targets in major cities—New York, Washington, Paris, London, Tokyo, and so on—have symbolic value simply because of their location.

- *Volatile assets:* Assets that, if bombed, could cause substantial secondary damage are potential targets. These include oil refineries, chemical plants, and nuclear power plants. Trucks and trains transporting poisonous chemicals are also potential targets; the results could be devastating if they were exploded in a crowded city.

- *Infrastructure:* Assets whose destruction can affect a large number of people are terrorist targets. These include power distribution centers, reservoirs and the water supply, shipping ports, and the like. The key here is how significant the effect is. There have been large-scale blackouts in the U.S. over the years and large phone system outages, but these are not "terror" in the same way that poisoning a city's water supply would be. Internet outages are not terror attacks; not being able to access your e-mail for a day is not terror, it's nuisance. And it isn't just central distribution points that are vulnerable. In 1984, followers of Bhagwan Shree Rajneesh infected restaurant salad bars in Oregon with salmonella, injuring 751. The postal system has been used as a distribution network for terrorist attacks: Both biological and conventional bombs have been sent through the mail.

- *Morale:* This is a nebulous asset, comprising people's perceptions of their own personal safety and their confidence in society. A terrorist wants people to feel fear every time they go shopping or every time they open a piece of mail. A terrorist wants a mother to keep her children home from school in Kansas because of a bombing in New York. The media is an important means for spreading the effects of a terrorist attack far and wide.

It's important not to lose sight of the forest for the trees. Countermeasures often focus on preventing particular terrorist acts against specific targets, but the scope of the assets that need to be protected encompasses all potential targets, and they all must be considered together. A terrorist's real target is morale, and he really doesn't care about one physical target versus another. We want to prevent terrorist acts *everywhere*, so countermeasures that simply move the threat around are of limited value. If, for example, we spend a lot of money defending our shopping malls, and bombings subsequently occur in crowded sports stadiums or movie theaters, we haven't really received any value from our countermeasures.

That being said, some assets deserve special protection. Airplanes are attractive terrorist targets because of their failure properties; even a small explosion aboard an airliner generally result in a crashed plane and the death of everyone onboard. Major sporting events are attractive not just because of crowd density, but because the death and destruction will be watched by millions on live television. (Security tip: If you are in a stadium when a bomb explodes, stay in your seat. You're more likely to be injured by a panicked crowd than you are by a terrorist.) Some targets are especially attractive because they encompass several categories. For example, blowing up a bridge or tunnel would kill many people, destroy a public symbol, and affect a city's infrastructure.

If the enormous number of potential targets seems daunting, it should be. One of the reasons terrorism is so hard to defend against is because there are just too many possible attacks.

Step 2: What are the risks to these assets? Terrorism is often viewed as a crime against people and property, but it's really a crime against the mind. The goal of a terrorist is to sow fear and terror. This goal is generally achieved through arbitrary, and seemingly random, death and destruction directed at innocents. Most of the damage in a terrorist attack isn't from the actual attack itself (as horrendous as that may be) but from the aftereffects: Everyone who watches the aftermath of the attack on television or reads about it in the newspaper becomes victimized, as well. In reality, there are three targets in any terrorist attack.

- *The physical target:* This is the school bus, or airplane, or shopping mall, or bridge. The target encompasses both the thing being attacked and any technological multipliers that might cause collateral damage (the release of chemical agents or toxic pollution, for example).
- *The economic target:* Imagine if a commercial airplane were shot out of the sky over Chicago by a terrorist with a shoulder-launched Stinger missile. (The U.S. gave many of these to the mujahadeen to fight the Soviets in Afghanistan, and those same weapons are now being sold on the arms black market.) In addi-

tion to the loss of life and property caused by the crash, the economic impact from that attack would be devastating. All planes might be grounded until a defense was deployed. Or people might be unwilling to fly—more so than after 9/11—effectively grounding and bankrupting the entire commercial aviation industry. Similarly, a terrorist attack at a major shipping port could shut down the nation's import infrastructure. Any attack of this nature would have enormous economic side effects, rippling out far beyond the cost of the incident itself.

- *The social target:* Fear can spread as quickly as a CNN broadcast. Panic is another potential problem: If a terrorist detonates a biological bomb in a city, more deaths could result from the ensuing panic than from the bomb itself. But panic happens more in movies than in real life; there was no panic after the Oklahoma City bombing and surprisingly little panic in Lower Manhattan after the 9/11 attacks. Perhaps a biological or nuclear bomb might result in some panic, but people are more likely to rise to the occasion.

We must take the economic and social components of any attack into account when we try to understand the threats. And an actual terrorist attack isn't even required: Even the vague threat of terrorism is enough to cause some fear and affect the economy. The first time the U.S. threat level was raised to Orange, the stock market declined sharply.

We also need to understand the attackers and their attacks. Most attackers, terrorists included, are copycat attackers. They're simply not innovative, and they're going to repeat the same attacks that have worked in the past—another bus, another restaurant—because they continue to be effective. Derailing a train isn't as popular as blowing up an airplane, even though it's far harder to defend against. Al Qaeda seems smarter than that; their attacks have been more innovative, but they're the rare exception.

Bombs are the quintessential terrorist tool, although there are examples of terrorists with automatic weapons opening fire in crowded places and killing people until they themselves are killed. (This kind of attack is more often perpetrated by murderous kooks, not terrorists.) There has been considerable discussion in the media about chemical, biological, and nuclear weapons that might fall into terrorist hands, but it is important to remember that terrorist weapons have generally been much more low tech and are likely to continue to be so.

Unfortunately, low tech doesn't mean ineffective. Some Palestinian suicide attacks have been remarkably deadly. Timothy McVeigh's bomb that destroyed the Oklahoma City Federal Building in 1995, killing 168, was a low-tech mixture of ammonium nitrate fertilizer and standard fuel oil, a combination that is used so commonly by terrorists that law enforcement has a nickname for it: ANFO. In February 2003, a fifty-six-year-old man with a history of mental illness set fire to a milk carton full of gasoline and tossed it into a crowded subway car in Taegu, South Korea; 196 people died in the ensuing station fire. This senseless tragedy wasn't terrorism, but it does demonstrate that very low-tech attacks can do a lot of damage. Weapons of mass destruction might grab the headlines, but milk cartons and fertilizer are the kind of weapons more commonly associated with terrorism.

Cyberterrorism has grabbed the headlines recently, but most of that is overblown, as well. We know what terrorism is. It's someone blowing himself up in a crowded restaurant or flying an airplane into a skyscraper. It's not infecting computers with viruses, disabling the Internet so people can't get their e-mail for a few hours, or shutting down a pager network for a day. That causes annoyance and irritation, not terror. This is a difficult message for some, because these days anyone who causes widespread damage is labeled "terrorist." But imagine for a minute the leadership of Al Qaeda sitting in a cave somewhere, plotting the next move in their jihad against the U.S. One of the leaders jumps up and exclaims: "I have an idea! We'll disable their e-mail. . . ."

Stories of terrorists controlling the power grid, or opening dams, or taking over the air traffic control network and colliding airplanes, are unrealistic scare stories. This kind of thing is surprisingly hard to do remotely. Insiders might have an easier time of it, but even then they can do more damage in person than over a computer network.

Listing possible threats is good for scaring people. It makes for compelling news broadcasts and exciting movie plots, but we need to make proper security trade-offs. We need to understand the actual risks. Here's the critical question we need to answer: Just how likely is a terrorist attack, and how damaging is it likely to be?

Terrorist attacks are very rare. So rare, in fact, that the odds of being the victim of a terrorist attack (in any industrialized country) are almost nonexistent. Modifying your life because of the risk is simply irrational. I know someone living in the DC area who refuses to fly or take public transportation because he fears a terrorist attack. But because automobile accidents are so common, he increases his overall risk of death significantly by driving instead. Several newspapers pointed out that by driving any more than a few miles to buy plastic sheeting and duct tape, Americans increased their overall risk more than if they had simply ignored that infamous advisory from the Department of Homeland Security.

Almost all terrorist attacks are minor affairs, affecting only a few people. Airplane bombings are the traditional exception, causing millions of dollars' worth of damage; the attacks of 9/11 were also an anomaly. Most terrorist attacks don't cause much damage directly. The real damage is from the secondary effects: people being too scared to fly or take public transportation. And while it's easy to imagine a disaster scenario or movie plot involving a nuclear bomb in terrorist hands, the reality is much more mundane. These kinds of attacks are very difficult to pull off, and very unlikely.

There is a real risk of terrorism, but the situation is not nearly as dire as most people thought in the months directly after 9/11. International terrorists are trying to attack the U.S. and Western Europe (and their interests around the world), but they're rare and they've become rarer since governments started counterattacking and arresting terrorists before they strike. It's impossible to quantify the risk in any meaningful sense, but we can look back at the previous decades and see the trends. International terrorism happens, but it's much less common than conventional crime.

None of this discussion is meant to belittle or deny the risks—it's just to put them in perspective. In 2001, 3,029 people died in the U.S. from terrorism (the 9/11 attacks). During that same year, 156,005 people died from lung cancer, 71,252 from diabetes, 41,967 from motor vehicle accidents, and 3,433 from malnutrition. Consider what we're willing to spend per year to cure diabetes or increase automobile safety, and compare that with the $34 billion we're spending to combat terrorism. The response to the terrorism threat has not been commensurate with the risk.

The problem lies in the fact that the threat—the potential damage—is enormous. Security is all about trade-offs, but when the stakes are considered infinitely high, the whole equation gets thrown out of kilter. In the frightened aftermath of 9/11, people said things like: "The budget for homeland security should be infinite. The trade-offs we need to make *should* be extreme. There's no room for failure." It was easy to succumb to hysteria and scare-mongering and to overre-

Deadliest Terrorist Strikes in the U.S.

Date	Attack and Location	Fatalities
11 Sep 2001	Crashing of hijacked planes into World Trade Center, New York, NY, Pentagon in Alexandria, VA, and site in PA	3,029
19 Apr 1995	Truck bombing of Federal Building, Oklahoma City, OK	169
16 Sep 1920	Bombing near bank in New York, NY	34
1 Oct 1910	Bombing at newspaper building in Los Angeles, CA	21
29 Dec 1975	Bombing at airport in New York, NY	11
23 Jul 1916	Bombing at parade in San Francisco, CA	10
4 May 1886	Bombing at Haymarket Square rally in Chicago, IL	7
26 Feb 1993	Truck bombing of World Trade Center, New York, NY	6

Significant Terrorist Acts Using Unconventional Weapons

Date	Attack and Location	Casualties
20 Mar 1995	Sarin nerve gas attack in Tokyo subway, Japan	12 killed, 5,511 injured
27 Jun 1994	Nerve gas attack in Matsumoto, Japan	7 killed, 270 injured
Sep–Oct 2001	Anthrax-laced letters to multiple locations in the U.S.	5 killed, 17 injured
19 Apr 1946	Cyanide poisoning in prison near Nuremberg, Germany	2,283 injured
15 Sep 1984	Salmonella poisoning in restaurants in The Dalles, OR, USA	751 injured
19 Apr 1995	Tear gas attack in Yokohama, Japan	400 injured

Source: "Worst Terrorist Strikes   U.S. and Worldwide," compiled by Wm. Robert Johnston, used by his permission.

Deadliest Terrorist Strikes Worldwide

Date	Attack and Location	Fatalities
11 Sep 2001	Crashing of hijacked planes into World Trade Center, New York, NY, Pentagon in Alexandria, VA, and site in PA	3,029
23 Jun 1985	Midair bombing of Air India flight off Ireland, and attempted bombing of second flight	331
8 Aug 1998	Truck bombings of U.S. embassies in Kenya and Tanzania	303
23 Oct 1983	Truck bombings of U.S. Marine and French barracks, Lebanon	301
21 Dec 1988	Midair bombing of Pan Am flight over Scotland	270
12 Mar 1993	15 bombings in Bombay, India	257
12 Oct 2002	Car bombing outside nightclub in Kuta, Indonesia	202
19 Sep 1989	Midair bombing of French UTA flight in Chad	171
26 Oct 2002	Hostage taking and attempted rescue in theater in Moscow, Russia (includes 41 terrorists killed)	170
19 Apr 1995	Truck bombing of Federal Building, Oklahoma City, OK	169
16 Apr 1925	Bombing of cathedral in Sophia, Bulgaria	160
14 May 1985	Armed attack on crowds in Anuradhapura, Sri Lanka *	150
3 Aug 1990	Armed attack at two mosques in Kathankudy, Sri Lanka *	140
2 Oct 1990	Crash of hijacked PRC airliner in Guangzhou, China	132
23 Nov 1996	Crash of hijacked Ethiopian Air flight off Comoros	127
18 Apr 1987	Roadway ambush in Sri Lanka *	127
13 Sep 1999	Bombing of apartment building in Moscow, Russia	124
13 Aug 1990	Armed attack at mosque in Eravur, Sri Lanka *	122
29 Nov 1987	Midair bombing of Korean Air flight near Burma	115
23 Sep 1983	Midair bombing of Gulf Air flight over the UAE	112
22 Sep 1993	Crash of airliner struck by missile in nation of Georgia *	106
21 Apr 1987	Bombing of bus depot in Columbo, Sri Lanka *	106
4 Dec 1977	Crash of hijacked Malaysian airliner, Malaysia	100
25 May 1973	Midair bombing of Aeroflot airliner, Siberia	100
13 Dec 1921	Bombing of Bolgard palace in Bessarabia (modern Moldova)	100

Note: Items marked with an asterisk are not usually considered to be terrorist attacks.

Source: "Worst Terrorist Strikes   U.S. and Worldwide," compiled by Wm. Robert Johnston, used by his permission.

act to the terrorist threat. In fact, one of the primary goals of terrorism is to create irrational terror far in excess of the actual risks. But this kind of talk is meaningless. When a country allocates an infinite budget to homeland security, the terrorists really have won.

Sensible security does not result from fear. Just because anomalies happen doesn't mean security has failed. The risk of a terrorist attack before 9/11 wasn't appreciably smaller than the risk of a terrorist attack after 9/11. Before 9/11, European countries mostly had an accurate assessment of their risks. In the U.S., the risks were largely underestimated; many people thought it couldn't happen there. But after 9/11, the risks in the U.S. suddenly became grossly overestimated. This situation has lessened somewhat, but the public perception of risk is still wildly out of proportion to the actual threat. The reality is that the risks are low; and even if some terrorist manages to set off a dirty nuke in the middle of a crowded city, the risks of an individual being affected by terrorism still won't change appreciably.

This is a precarious position to take politically, which is why I believe most politicians have steered clear of it. It's safe for a political leader to make dire predictions about the future and recommend an extreme course of action. If another terrorist attack happens, then she can say that the event proved her right. And if nothing happens, she can claim that her security program was a success. (And that it keeps away the vicious purple dragons, too.) A politician is on shaky ground when he says, "Don't worry; it's not that bad." He looks ineffectual compared to his colleagues who are trying to *do something* to make us safer (even if the "something" doesn't really make us safer); worse, he looks as if he doesn't care about his constituents. And if another attack happens, he looks even worse. The public's difficulty in assessing risks plays into this. Many people have been frightened into believing that terrorism is a far greater risk than it is. All sorts of incidents are immediately assumed to be terrorism, even though investigations prove other causes. And when the next terrorist attack occurs on U.S. soil, people, politicians, and the press will all exaggerate the risks of terrorism even more.

Here's the bottom line when you realistically and unemotionally assess the risk to your personal security of a terrorist attack: If you're don't live in a major coastal metropolitan city or next to a nuclear power plant or chemical factory, you're more likely to die of a bee sting than a terrorist attack. Even if you do live in a big city or next door to a power plant, the odds of being a terrorist victim are still vanishingly small. Any precautions you take should be directed toward and in proportion to those risks.

Step 3: How well does the security solution mitigate those risks? There are exceptions, but most of the countermeasures put in place after 9/11 don't do a very good job of mitigating the risk of a terrorist attack. Many of the problems are inherent in the characteristics of the assets and the risks to those assets.

- Defending targets is hard. It's one thing to defend a military target—or even a military office building like the Pentagon—and quite another to defend the sorts of civilian targets that can be attacked. In military parlance, most of the assets we need to defend are known as soft targets: undefended nonmilitary sites. These targets are meant to be accessed all the time by all sorts of people; they have poorly defined perimeters, and they're generally situated in populated areas. These characteristics make them particularly hard to defend.
- Preventive countermeasures are largely ineffective because of the huge number of possible targets. Defenses that simply shift the attacks from one target to another are wasteful, although—as I said before—it is important to defend particular high-profile targets nonetheless.
- Detection and response is more effective than prevention. The notion of secure preventive barriers around many of these targets is simply nonsensical. There's simply no way to keep terrorists off buses, weapons off airplanes, or bombs out of crowded intersections. Far more effective is to detect terrorist attacks in the planning stages and to respond before damage can occur. Of course, doing this is very difficult, as well; most terrorist attacks happen too quickly for defenders to respond before the damage is done. Detection and response are more useful for mitigating the effects of an attack than they are for preventing it from happening.
- Benefit denial is a critical countermeasure. Morale is the most significant terrorist target. By refusing to be scared, by refusing to overreact, and by refusing to publicize terrorist attacks endlessly in the media, we limit the effectiveness of terrorist attacks. Through the long spate of IRA bombings in England and Northern Ireland in the 1970s and 1980s, the press understood that the terrorists *wanted* the British government to overreact, and praised their restraint. The U.S. press demonstrated no such understanding in

the months after 9/11 and made it easier for the U.S. government to overreact.

- Proactive countermeasures, such as advance detection and counterattack, are more effective than reaction. In general, reactive countermeasures are only a small part of the answer; we will never be able to stop terrorism solely by defending the targets. The only way to deal effectively with terrorists is to detect terrorist plots before they're implemented, then counterattack and go after the terrorists themselves: rolling up terrorist networks, disrupting funding streams and communications, and so on.

- Long-term countermeasures, such as deterrence and education, are the only real solution, and even they are imperfect. The best way to reduce terrorism is to solve the underlying socioeconomic and geopolitical problems that cause it to arise in the first place. This isn't absolute, but nothing in security is ever absolute. It is also extremely difficult, and some of the problems seem, and may be, impossible to solve. Deterrence has a place, as well, although it is unclear how effective it is against suicide terrorists.

To summarize: Prevention is impossible. Mitigation is important. Intelligence and counterattack are critical. And none of this is as effective as addressing the root causes of terrorism.

Step 4: What other risks does the security solution cause? One risk of the post-9/11 countermeasures is crime. Terror attacks are rare, while crime is common. Countermeasures that mitigate rare terrorist attacks but amplify common crime are, on the whole, damaging to society. I'll give three examples. One, the current system for passenger screening at airports makes it easier for thieves to steal laptop computers by distracting passengers while simultaneously separating them from their laptops. Two, corporate officers committing fraud, embezzlement, and regulatory violations can hide behind the secrecy provisions of some antiterrorism countermeasures. (If a chemical company's inner workings are secret, you might never learn that it's polluting the river.) And three, gathering information about all of us into large surveillance databases makes frauds involving identity theft easier.

A much greater risk is embodied in the trusted people necessary to implement the security systems themselves: the law-enforcement officers, the data-entry clerks, the computer technicians, and legions of bureaucrats. Many of the laws passed in the U.S. to help fight terrorism, most notably the USA PATRIOT Act, give the government broader powers of surveillance and spying, and more leeway to investigate, arrest, and detain suspects (both citizens and noncitizens) without formal charges and with less judicial oversight. All of this is fine if we assume that the government is completely benevolent and that these powers will be used only for good, but history has taught us again and again that power is a corrupting influence. The problem with giving powers like this to the state is that they are far more likely to be used for the benefit of those in power than to protect citizens, and the reason the U.S. Constitution and the court system have put limits on police power is that these limits make all citizens more secure. We're more secure as a society because the police have limited powers. We're more secure as a society because we have the right to free speech—even the right to criticize the very government that gives us that right—and the right to freedom of the press. The framers of the U.S. Constitution adopted some novel countermeasures to defend what they thought of as very vulnerable freedoms against many threats, of which the most critical were people attempting to gain dictatorial or monarchic powers or exercise a "tyranny of the majority." Over 220 years, their countermeasures have been so effective and have required so little modification that we often take them for granted, but that doesn't mean the threat is gone. Constitutionally protected liberties are more important to individual security than are democratic elections, and taking away liberties in the name of security is an enormous trade-off. As Benjamin Franklin said in 1784: "They that give up essential liberty to obtain a little temporary safety deserve neither liberty nor safety."

Step 5: What trade-offs does the security solution require? Many of the countermeasures proposed have enormous trade-offs. Total monetary costs for the U.S. war on terror are over $40 billion per year, and that doesn't include the war with Iraq and its aftermath. And the social costs for these countermeasures are enormous: in liberty, privacy, time, convenience, and the economic fallout of these costs in loss of business and market confidence. This book isn't about the value of these things, but with any security countermeasure, it is important to look carefully at what trade-offs that countermeasure requires and to keep several things in mind:

- Agendas abound. Often the players implementing the security countermeasures aren't the ones making the trade-offs. It's easy for a third-generation American to back countermeasures that require

the investigation of all Iraqi immigrants, for example. "No fly" lists
of people who are not allowed to board airplanes might seem like
a good idea in general, unless you're one of the people who have
been mistakenly put on the list and then learned that there's no
procedure for getting yourself off.

- Liberties are far easier to give away than to get back. Extreme
  trade-offs made in the emotional aftermath of 9/11 will be with us
  for years, if not decades. Don't give up liberties lightly.
- Countermeasures that permeate every area of society are expensive. It's important to understand the exact monetary costs of any
  countermeasure before deciding whether it's worth it. And like all
  large infrastructure projects, the costs to operate and maintain
  these systems are significantly larger than the costs to develop and
  deploy them.

And that's the big question: Is a particular countermeasure worth
it? The answer depends on the details of the countermeasure, but here
are some general principles that can guide specific determinations:

- Humans provide better security than machines. Intelligent and
  trained vigilant guards are an excellent countermeasure, but not
  for anything other than identifying high-profile terrorist targets.
  Guards can be trained to know what to look for, and their presence increases the likelihood that the attackers will slip up and
  make a mistake. Guards can also respond to new attacks and
  unanticipated suspicious activity; they can respond to the unexpected. But there's a cycle of diminishing returns to watch out for:
  doubling the guards doesn't double security, and there will be a
  point where adding more guards won't make any appreciable security difference. There will also be a point where guarding more
  targets doesn't help, because of the sheer number of potential targets. And as in all antiterrorist countermeasures, the active failures
  will far outnumber the passive failures. But used sparingly,
  increased police presence—both uniformed and undercover—
  works.
- Simple is better than complex. In the months after 9/11, people
  discussed systems to detect nuclear material on roads around large
  cities or at ports automatically. While research into this kind of
  potential threat should continue, the enormous expense of implementing such a system, the prevalence of false alarms (which leads

to the "Boy Who Cried Wolf" syndrome of ignoring all alarms), and the ease with which a smart attacker could bypass an automated system demonstrate that this sort of added security is rarely worth the trade-offs. Spending more money on intelligence and investigation is far more cost-effective, because it targets the attackers, rather than waiting for the attackers to come to the defensive systems.

- Freedoms and liberties actually provide security. People living in open societies like the U.S. and the European Union are more secure—as a whole—than people living in countries where surveillance and restrictions on personal actions are commonplace and liberty and the rule of law are weak: the former Soviet Union and many Warsaw Pact countries, China, North Korea, and much of Africa and the Middle East. It seems paradoxical, but societies that have the most restrictive regulations hampering law enforcement are the safest societies. And countries that allow freedom of thought and expression are the ones that come up with innovative defensive ideas and new defensive technologies. World politics demonstrates this fact again and again. Even if we gave up our freedoms, we wouldn't be more secure as a result.

．．．．

With this general discussion of the five steps in mind, we can look at particular countermeasures and judge whether they're worth it. But first, let me ask you a question: If I offered to sell you a device that would substantially increase the odds of your surviving an automobile accident, you'd be interested, right? But what if that device added $10,000 to the cost of your car? Still interested? Maybe not. If the cost were $1,000? Maybe—antilock breaks and airbags are common now. This is what we have to look at with the issue of national security: the trade-offs.

A complete and thorough examination at a security boundary is rarely an effective enough security countermeasure to be worth the trade-offs. It's simply not possible to search every container of goods at every port in the U.S., even though it is possible that one of them may, just may, contain a nuclear or biological terrorist weapon. The borders are too big, and too many goods—$1.1 trillion in 2001—are coming into the country. Even trying would become cost-prohibitive very quickly. In one month alone, the U.S. imported over a billion dollars' worth of shoes; who would we get to look into every shoe box?

Similarly, screening all checked baggage at airports, even though recently implemented in the U.S., provides minimal security at a very high cost. X-raying every piece of carry-on luggage and forcing every passenger to walk through a metal detector is worth the trade-offs only because airplanes are a higher-profile target than average; similar security measures also might be worth it at a courthouse, but the same countermeasures simply aren't worth it for an average office building.

Random and profiled screening of both people and transported goods is only marginally effective, but a good idea nonetheless. We can't search everything, but we should search a small percentage. Even the possibility of detection puts a plan in doubt and increases the likelihood that the attackers will make a mistake. Search criteria should be a combination of random searching and profiling, much like what occurs on an international border today. Care must be taken to ensure that profiling criteria do not unduly violate civil liberties, both because the potential trade-off to dignity and societal values is significant and because simple racial profiling is ineffective. Note, however, that this countermeasure works only when the number of attackers is relatively small. Screening hasn't stopped the flow of illegal drugs into the country, simply because there are just too many smugglers and the police are too overwhelmed to backtrack every arrested smuggler to his source. We can't afford to be that lackadaisical about terrorism.

Many of the new airport security countermeasures are ineffective and too expensive. Photo ID requirements are ineffective. Passenger profiling is marginally effective, but massive data mining systems like CAPPS-II are not worth it. X-ray machines and metal detectors are a good defense, but smart terrorists can cheat and bypass them. Hand-searching selected carry-on baggage is marginally effective; the random detection system means that a terrorist can't count on bringing a weapon on an airplane, but far more things can be used as weapons than are confiscated at checkpoints. National Guard troops at security checkpoints are ineffective, but probably were worth it in the first months after 9/11 because they reassured many people (defending morale, not material). Positive bag matching is effective against non-suicide bombers only. Many countermeasures defend against dumb terrorists and copycats. There really isn't anything that can defend against a smart and sufficiently motivated terrorist. (Remember that the 9/11 terrorists took practice flights to see how airline security worked and how it could best be bypassed.) When you examine the details, only two effective antiterrorism countermeasures were taken in the wake of 9/11: strengthening cockpit doors and passengers learning they need to fight back. Everything else—let me repeat that: *everything else*—was only minimally effective, at best, and not worth the trade-offs.

The Terrorist Information and Prevention (TIPS) program proposed by the U.S. government in 2002 was badly flawed. This was the system by which people who regularly interacted with many members of the public—meter readers who enter people's homes, for example—would report on any suspicious activity. Thankfully, this program was killed. It would have been swamped by false alarms and questionable data, probably would never have uncovered any terrorist cells, and would require enormous trade-offs in liberties and privacy. This kind of state-sponsored mass snitchery would be no more effective in the U.S. than it was in East Germany or the Soviet Union.

Financial tracking of terrorist organizations is a good idea, but broad surveillance of money movements in an effort to find terrorists is expensive and ineffective. It's the same false alarm problem; terrorists are rare, and they're going to make efforts to hide their financial dealings. A system like this is far more likely to catch innocents. Still, to the extent we can freeze terrorist assets, we should do so. Unlike defending targets, which only forces terrorists to change tactics, eliminating terrorist financing has lasting effects regardless of the target. On the other hand, terrorism can be surprisingly inexpensive.

The color-coded threat alerts issued by the Department of Homeland Security are useless today, but may become useful in the future. The U.S. military has a similar system; DEFCON 1–5 corresponds to the five threat alerts levels: Green, Blue, Yellow, Orange, and Red. The difference is that the DEFCON system is tied to particular procedures; military units have specific actions they need to perform every time the DEFCON level goes up or down. The color-alert system, on the other hand, is not tied to any specific actions. People are left to worry, or are given nonsensical instructions to buy plastic sheeting and duct tape. Even local police departments and government organizations largely have no idea what to do when the threat level changes. The threat levels actually do more harm than good, by needlessly creating fear and confusion (which is an objective of terrorists) and anesthetizing people to future alerts and warnings. If the color-alert system became something better defined, so that people know exactly what caused the levels to change, what the change means, and what actions they need to take in the event of a change, then it could be useful. But even then, the real measure of effectiveness is in the implementation. Terrorist attacks are rare, and if the color-threat level changes willy-nilly with no obvious cause or effect, then people will simply stop paying attention. And the threat levels are publicly known, so any terrorist with a lick of sense will simply wait until the threat level goes down.

Stockpiling vaccines in cities sounds like a good precaution, but it would be mostly ineffective in the event of a biological terrorist attack. After an attack, there is rarely time to distribute the vaccine. Vaccinating emergency personnel as a precaution is a more complicated decision. On one hand, making vaccinations available free to emergency personnel both provides some protection in the event of an attack and reduces anxiety for the affected workers. On the other hand, it's largely security theater, and every vaccine has its own—albeit rare— side effects: A couple of people with heart problems have died as a result of smallpox vaccinations in the U.S.

Massive surveillance systems that deprive people of liberty and invade their privacy are never worth it. By these I mean systems that are put in place as precautions, and not dragnets used after an event to ensnare attackers. The latter can work well, but for the former the costs are too high, both social and economic, and the benefits are negligible.

In general, the costs of counterterrorism are simply too great for the security we're getting in return, and the risks don't warrant the extreme trade-offs we've been asked to make. I understand that this is a subjective and personal analysis, based on how much risk people are willing to accept. But looking at the risks and trade-offs rationally, the analysis isn't very promising.

The organizations chartered with the mission of preventing terrorism have been given an impossible job. There's no way to prevent all future terrorist attacks. Terrorism will continue for the foreseeable future, as it has throughout history. The best we can hope for is to mitigate the risks without too much in the nature of onerous trade-offs. Prevention of terrorism is no different from any other aspect of security.

····

Ironically, in the two years since 9/11, we've got the security level mostly right but the costs wildly wrong. The security we're getting against terrorism is largely ineffective, although it's probably commensurate with the minimal level of risk that actually exists. But it comes at an enormous expense, both monetarily and in loss of privacy.

To understand why people were willing to give up their privacy to attain a feeling of security, regardless of how effective that security actually was, you need to recall the general mind-set in the months after 9/11. In the aftermath of the mind-numbing shock and horror, people needed to do something immediate, and invasive countermeasures seemed the easiest solution. Many Americans declared themselves willing to give up privacy and other civil liberties in the name of security. They declared it so loudly that this trade-off now seems to be a *fait accompli*. Pundit after pundit has talked about the balance between privacy and security, discussing whether various increases of security are worth the privacy and civil liberty losses. This discussion seems odd to me, because linking the two is just plain wrong.

Security and privacy, or security and liberty, are not two sides of a teeter-totter. This association is both simplistic and misleading. Security is always a trade-off, yes, but privacy and liberty are not always the things traded off. It's easy and fast, but not very cost-effective, to increase security by taking away privacy. However, the best ways to increase security are not necessarily at the expense of privacy or liberty. Use airline security as an example: Arming pilots, reinforcing cockpit doors, and teaching flight attendants karate are all examples of security measures that have no effect on individual privacy or liberties. Other effective countermeasures might be better authentication of airport maintenance workers, panic buttons and automatic controls that force planes to land automatically at the closest airport, and armed air marshals traveling on flights.

And privacy- or liberty-reducing countermeasures often require the most onerous trade-offs. At this writing, the U.S. has arrested about a thousand people and is holding them incommunicado, without allowing them trials or hearings or, in many cases, access to an attorney. It's likely that among these people are a few would-be terrorists, and keeping them in jail has probably been pretty effective at preventing further terrorist attacks. But the cost—arresting a thousand innocent people—has been enormous.

Lack of planning is why we saw so much liberty-depriving security directly after 9/11. Most of the assets that had to be defended were not designed or built with security in mind. Our leaders had an extreme reaction driven by a grave perceived risk, and they wanted a lot more security—quickly. People in power were asked "What do you need to fight a war on terror?" There was no time to think through security and choose countermeasures based on their effectiveness and trade-offs. Nearly all the proposals were things that the FBI, the CIA, and various factions within the administration had been wanting for a long time. "Give us more money" and "Give us more power" were natural answers that required very little detailed analysis.

That analysis needed to come from outside the FBI, and outside the administration, but no one was willing to do it so soon after 9/11. The most politically expedient option was to slap highly invasive and expensive countermeasures on top of existing systems. At the same time, people wanted to be reassured, responding more to the feeling of security than to the reality; and because they were driven by fear, they accepted countermeasures that required extreme trade-offs. People felt that they must be getting *something* because they were giving up so much.

That was two years ago, though, and it's about time we replaced these invasive systems with good security that mitigates the real threats while minimizing the necessary trade-offs. It's about time we made sensible security trade-offs.

· · · ·

Which brings us to the Department of Homeland Security. According to its own writings, the "mission of the Department of Homeland Security is to: (1) prevent terrorist attacks within the U.S.; (2) reduce America's vulnerability to terrorism; and (3) minimize the damage and recover from attacks that do occur." The most heartening thing about this mission statement is the third item: the open admission that absolute success is impossible and that there are limits on the department's preventive abilities.

Unfortunately, the Department of Homeland Security is far more likely to increase the country's vulnerability to terrorism. Centralizing security responsibilities will create a commonality of approach and a uniformity of thinking; security will become more brittle. Unless the new department distributes security responsibility even as it centralizes coordination, it won't improve the nation's security.

The dual requirements that security decisions need to be made as close to the problem as possible, and that security analysis needs to happen as far away from the sources as possible make the problem subtle. Security works better if it is centrally coordinated but implemented in a distributed manner. We're more secure if every government agency implements its own security, within the context of its department, with different individual strengths and weaknesses. Our security is stronger if multiple departments overlap each other. To this end, it is a good thing that the institutions best funded and equipped to defend our nation against terrorism—the FBI, the CIA, and the military's intelligence organizations—aren't part of this new department.

But all these organizations have to communicate with each other, and that's the primary value of a Department of Homeland Security. One organization needs to be a single point for coordination and analysis of terrorist threats and responses. One organization needs to see the big picture and make decisions and set policies based on it.

Your body doesn't have a Center for Body Security; fighting disease is a function distributed throughout every organ and every cell. You have all sorts of security systems, ranging from your skin which keeps harmful things out of your body, to your liver filtering harmful things from your bloodstream, to defenses in your digestive system, to your immune system fighting off diseases. These systems all do their own thing in their own way. They overlap each other, and to a certain extent one can compensate when another fails. It might seem redundant and inefficient, but it's more resilient, reliable, and secure. You're alive and reading this book because of it.

The biological metaphor is very apt. Terrorism is hard to defend against because it subverts our institutions and turns our own freedoms and capabilities against us. It invades our society, festers and grows, and then attacks. It's hard to fight, in the same way that it is hard for the body to fight cancer. If we are to best defend ourselves against terrorism, security needs to be pervasive. Security needs to be everywhere. Every federal department needs to do its part to secure our nation. Fighting terrorism requires defense in depth, which means overlapping responsibilities to reduce single points of failure, both for actual defensive measures and for intelligence functions.

The U.S. would be less secure if the new Department of Homeland Security took over all security responsibility from the other departments. It would be a security disaster for the Department of Energy, the Department of Commerce, and the Department of State to say "Security? That's the responsibility of the Department of Homeland Security." Security is the responsibility of everyone in government. Terrorism can't be defended against with a single countermeasure that works all the time. Terrorism can be defended against only when every countermeasure works in its own way and together with others provides an immune system for our society. The Department of Homeland Security needs to coordinate but not subsume.

Anyone implementing a counterterrorism security system had best learn humility. It's impossible to prevent all terrorist attacks. It's impossible to eradicate terrorism from the planet; recovering from and minimizing the damage from terrorist attacks form important parts of defense. Teaching people to remain calm, and not to live in fear, is another part of defense. Refusing to overreact after a terrorist attack is another. We *can* win the war on terror using sensible security, if we can convince people that sensible security is in fact the solution.

Example: Terrorist Information Awareness

In 2002, the U.S. government embarked on an experimental data mining program designed to sweep through large swaths of data looking for suspicious things that indicated terrorist activity. This predictive countermeasure is called Terrorist Information Awareness (TIA); the acronym originally stood for Total Information Awareness. There was public outcry against the system, and Congress approved funding for research but not for deployment against American citizens. Certainly this sort of thing will resurface, so let's walk through the five steps with the idea.

Step 1: What assets are you trying to protect? Potentially, everything.

Step 2: What are the risks to those assets? Terrorism. It's hard to be more specific in this case, because the TIA database proposal is more a response to terrorism angst than a response to a specific threat.

Step 3: How well does the security solution mitigate those risks? Not well. Predictive data mining systems are largely ineffective. The problems are twofold:

One, as long as there's no clear delineation of what to look for, the systems won't be able to find it. Before 9/11, individual suspicious people enrolling in flight schools wouldn't have raised an alarm. Today, they would raise an alarm even without such a system. We have no idea what Al Qaeda's next plot is, but I can guarantee that it won't raise an alarm.

Two, the false alarm rate would overwhelm the system. Even a problem as straightforward as picking terrorists out of crowds with face-recognition software has so many false alarms as to be useless; TIA would be even worse. Terrorists are rare, and a system that spits out 100,000 false alarms, each one of which needs to be investigated manually, for every real attack would rapidly overwhelm police.

Step 4: What other risks does the security solution cause? The databases themselves would become prime targets for criminals, both outsiders wanting to break into the system and steal the data and insiders already trusted with the data. Common criminals are far more numerous than terrorists, and the most secure solution is to not aggregate this data in the first place.

Step 5: What trade-offs does the security solution require? The monetary cost for these sorts of systems would be astronomical. It's not the algorithms that sift through the data looking for patterns that would cost so much but getting the data in the first place. The systems that need to provide input into the data mine are many and varied, and largely incompatible with each other. The FBI reports enormous problems getting its own internal systems to share information, and that's an example where one organization is in charge of every aspect of the problem and solution.

There are also the enormous privacy and liberty concerns surrounding a massive government database on everyone and everything.

TIA is not worth it. It doesn't even come close. Someday artificial intelligence might reach a point where these kinds of systems might be considered, but we're nowhere near there yet. Implementing a system like this would force us to make enormous trade-offs in return for minimal additional security.

Some have defended this program as being necessary for evidence collection after a terrorist attack. That argument makes even less sense; the police already know how to collect evidence from diverse sources, and they demonstrated how effective they can be in the days and weeks following 9/11. Giving them a less effective and much more expensive way of doing what they already do well is simply a bad trade-off.

Part Three

The Game of Security

Chapter 16

Negotiating for Security

The five-step process is useful for evaluating personal security decisions, but many security decisions involve a variety of players—each with his own agenda. Your ability to take control of important security decisions is often severely limited, but you do have some control. As a citizen, you can effect some changes in security practices with your vote. As a consumer, you effect others with your wallet. As a technologist, you can invent something that changes security. And if you have some measure of money and freedom, you can change your environment, even if that means relocating yourself. Making changes in security arrangements is also achieved through negotiation. Using mechanisms like insurance and laws that enforce liability, individuals can secure some measure of power in determining what kind of security will be available to them.

I started this book saying that people make security trade-offs based on their individual agendas, both for security and non-security decisions. When faced with a security countermeasure, you have to evaluate its effectiveness in mitigating *your personal risk in your personal situation*, and then you have to determine what the trade-offs are and if they're worth it *to you*.

The five-step process is designed to focus on the specific aspects of security you need to understand in order to make one basic decision: Is the security countermeasure worth the trade-offs? Again, here are the steps:

- *Step 1: What assets are you trying to protect?* Answering this question is essential because it defines the system under consideration. So much of the bad security surrounding us is a result of not understanding exactly what is being protected and of implementing countermeasures that move the risk around but don't actually mitigate it. And remember, often it's not simply a set of physical assets that are important, but particular functionalities of those assets. The assets that need securing are really a system, and you won't be able to protect them unless you understand what they are, how they work, and what aspects of them the attackers are after and why. (See Chapter 4.)

- *Step 2: What are the risks against these assets?* Answering this question means understanding the possible threats against the assets. Understanding this, in turn, involves analyzing the attackers and their goals and, finally, the attacks they might launch to achieve those goals. A full understanding of the risks requires determining how likely the various threats are, as well as their ramifications. Answering this question also requires evaluating how technological advances might affect potential attacks and attackers, and how that in turn might affect the risks. (See Chapters 5–7.)
- *Step 3: How well does the security solution mitigate the risks?* Answering this question requires an understanding of how the security countermeasure protects the assets against the risks and, more important, what happens when the security solution fails. As we've seen, answering this question can be very complicated. A countermeasure can mitigate the risk completely, partially, or not at all. A countermeasure can be more effective against one particular attack (or one particular type of attacker) and less effective against another. A countermeasure can fail both passively, by allowing an attack, and actively, by blocking legitimate access to the assets being defended. Being able to answer this question well means you're getting at the heart of security. (See Chapters 8–15.)
- *Steps 4: What other risks does the security solution cause?* Answering this question requires you to understand how the countermeasure interacts with other countermeasures, and how the security countermeasure works within the context of the overall system in which it is embedded. Almost all security countermeasures cause additional security risks, and it is vital to understand what they are. (See Chapters 8–15.)
- *Step 5: What trade-offs does the security solution require?* Answering this question requires you to understand how the countermeasure interacts with everything else: with all of the non-security components of the system. All countermeasures affect the functionality of the assets being protected. All countermeasures affect other systems. All countermeasures have a cost: not necessarily financial, but in terms of convenience, usability, freedoms, and so on. These trade-offs may have nothing to do with security, but often they are more important than security. (See Chapters 2, 3, and this chapter.)

After going through this five-step process, you can determine if the security countermeasure is worth it—not in some abstract sense, but practically and personally to you. You can determine if the trade-offs—money, inconvenience, loss of liberty, or any problems caused by the security countermeasure—are worth the additional security the countermeasure gives you with respect to your perception of risk in your personal situation.

Sometimes the answer is obvious. In Chapter 6, I analyzed the choice of not sending credit card numbers over the Internet. By the third step, it was clear that the countermeasure didn't mitigate the real security risk. By the fifth step, it became obvious that refusing to send credit card numbers over the Internet denied people the convenience of buying things online. Conclusion: It makes no sense to avoid sending credit card numbers over the Internet simply because they might be stolen. The risk is simply not that great, and the trade-off is too onerous for the minimal risk.

Often the answer is subjective, and very personal. In Chapter 8, I analyzed wearing a money belt while traveling. I went through the five steps and personally decided that the countermeasure had some benefit without any onerous trade-offs. But even so, I generally do not wear one. I decided that the additional security is not worth the bother, minimal as it is. Of course, I know other travelers who wear money belts every time they go abroad; to them, the additional security is worth it. Every traveler gets to decide for himself.

Sometimes the answer is complex. In my discussions about counterfeiting, I said that each year the U.S. spends more money securing its currency against counterfeiting, and tracking down counterfeiters, than the total annual value of existing counterfeiting. On the face of it, this seems irrational. But even if an initial examination of the problem indicates that the countermeasure isn't worth it, a more detailed analysis may very well conclude that unless the U.S. keeps the amount of counterfeiting low, the damage would quickly spiral out of control and become enormous. In this case, it makes sense to spend more to mitigate the risk than the actual current damage because the potential damage is so great.

And far too often the answer is unknowable. The five-step process requires that you assess the trade-offs, which implies that you can measure the risk accurately—and there are times when you just can't. For example, it is impossible to judge the efficacy of countermeasures against a threat that cannot even be imagined: alien invasion, for example. Additionally, it's sometimes impossible to get accurate data on easily imaginable threats. While neighborhood crime rates are generally public and more or less knowable, it's another thing entirely to know how effective a particular home burglar alarm system will be in deterring burglars. When attacks are rare, there just isn't enough data for accurate statistics, and attackers don't sit down to answer market surveys. In these situations, the best we can do is make educated guesses.

Determining exactly what to do, and what kind of countermeasures to deploy, might involve many iterations through the five steps. Imagine you're a homeowner, and you're concerned about burglaries. Should you have a door lock? You go through the five steps, and perhaps the problems resulting from forgetting to carry your key are not worth locking the door. But the countermeasure has lots of variants. What if you lock your door, but hide a key in the rafters of your garage just in case you forget yours? Now you go through the five steps again, and the cost–benefit trade-off is different. There's a little less security than if you didn't have an extra key in the garage, but there are fewer functionality problems. If you read an advertisement about a cheaper lock, you have to go through the five steps again. Is the lower cost worth the lowered security, or is the security of the two locks basically the same? And so on. The final result can be counterintuitive: A security countermeasure that is very effective might require too many trade-offs to be worthwhile, while a marginally effective security countermeasure might be so easy, cheap, and painless to implement that it would be foolish not to use. But you'll never learn that if you don't go through the five steps again and again.

And the first acceptable trade-off you find is probably not the best trade-off; you need to reexamine the countermeasure and see if there is a better way of doing the same thing. Maybe with a little more engineering you can design a system that provides the same amount of security with less of a privacy trade-off, or less of a convenience trade-off. Far too often I see people stop this process as soon as they find one system that works; modern security systems are complex enough that getting to an optimal result may require many, many iterations through this process.

Of course, that result won't remain optimal forever; you'll revisit it again and again over the years. You may modify your answer over time, based on new information about risks, or new realizations about what trade-offs you're willing to accept, or new technology that becomes available, or changed financial circumstances or ... the list goes on. The point is, security is never done; it's a never-ending process. You might add a second door lock if the crime rate in your neighborhood goes up or when you have a more expensive television in your house. Perhaps you go through the five steps and realize that no lock is strong enough to mitigate the risk sufficiently, so you start looking at alarm systems. You might decide they're too expensive, or you might decide they're worth it. Perhaps you initially decide they're too expensive, but a few years later wireless systems become cheap enough to make you change your mind. Perhaps turning the alarm system on every day is too much trouble, so you activate it only when you're away from home or away on vacation. Then perhaps you decide that the crime rate is rising even more, so you use the alarm every day. Or perhaps the crime rate falls, and you don't bother turning it on at all. And so on.

To some, this iteration process may feel unnatural. Why not build it strong, secure, and correct, once and for all ... and then forget about it? We might intuitively feel more comfortable with that kind of massive static fortress approach, but it's a trap. That approach is how you end up with bad security trade-offs and security that fails against smart attackers. Both risks and trade-offs are constantly in flux, and constantly reviewing and rethinking your security choices is your best hope for staying ahead of the attackers.

The best security systems evolve, so constant reevaluation and updating are essential for improving security, experimenting with new security, and replacing stopgap security with more robust systems. In the weeks and months after 9/11, the U.S. implemented all sorts of new laws and countermeasures designed to combat terrorism. Most of them weren't nearly as effective as they initially were thought to be, but the trade-offs were believed to be worth it nonetheless because people desperately needed a feeling of security. Two years later, some countermeasures were no longer in force; others evolved to take their place. And years from now, when we feel safer again, many of the countermeasures we're taking for granted now will no longer be around. More security and then less security and then more security, back and forth ... it's a process without end.

Understanding how this process works in practice involves understanding the various players and their agendas. In the homeowner example in Chapter 12, one person was in charge of her security so the security decision was hers alone to make. But rarely do we stand naked and alone facing a charging tiger, deciding whether to fight or run. Every player involved in a security decision has his own subjective perceptions of the risks and his own agenda of what he is willing to trade off to mitigate those risks. All players try to make the results of the security decision conform as closely to their ideal as possible by negotiating with the other players. In any system, there are players who pay the cost for security and players who reap the benefits. There are players who shoulder the risk and players who, for whatever reason, actually prefer lax security.

Modern security often involves multiple players and multiple agendas. In these situations, you might need to convince some other player to do something in order to improve your security, or you might be a hapless bystander as various other players negotiate your level of security and the trade-offs required of you. In short, you're going to have to negotiate.

The results of any negotiation depend more on the relative power of the players than on anything else. Back in Chapter 3, I gave the hypothetical example of all the players involved in airline security and their agendas. The public, the airlines, the pilots, the flight attendants, government officials, and the FAA were all involved in the negotiations, but not everyone's voices were heard to the same degree. After 9/11, the government wanted to ban laptop computers, but the airlines screamed that doing so would enrage their highest-revenue-generating passengers: business travelers. The tobacco lobby even got involved, ensuring that smokers would still be able to carry their matches and lighters aboard airplanes.

The one thing missing from the negotiations was us, the customers. The government—in particular, the FAA—was our proxy in these negotiations, but our voice wasn't heard as well as it might have been because the FAA serves many masters and often is more concerned with consensus than with being a perfect proxy. Decisions about airline security, decisions that affect us every time we fly, were made largely without us.

This is an important point. Because we—that is, you and I, as ordinary citizens—have so little power as individuals, we have almost no control over most of the major security systems that affect our lives. In the example above, the most effective way you can change your security is to opt out: You can either fly or not. Even in the area of home security, where you seem to have much more control, most of your perceived power is illusory. If you're trying to secure your home against burglary, it is true you have a few real options, but basically you can either live with the level of security in your neighborhood or move. You can install legal countermeasures, such as hard-to-pick locks, strong doors, bars on your windows, and an alarm system. Some countermeasures are prohibited by law; it's illegal for you to put a minefield in your yard. If you're rich, you can hire guards to patrol your estate and monitor the security cameras at its perimeter. But honestly, none of these things will increase your security as much as moving to a neighborhood with a lower crime rate. And if you're a resident of Kandahar, the West Bank, or a U.S. inner city slum—that is, either poor or living under a repressive government or both—you may not even have that option.

Your security largely depends on the actions of others and the environment you're in. For example, how tamper-resistant food packaging is depends more on government packaging regulations than on your purchasing choices. The security of a letter mailed to a friend depends more on the ethics of the workers who handle it than on the brand of envelope you choose to use. How safe an airplane is from being blown up has little to do with your actions at the airport and while on the plane. (Shoe-bomber Richard Reid was the rare exception to this.) The security of the money in your bank account, the crime rate in your neighborhood, and the honesty and integrity of your police department are out of your direct control. You simply don't have enough power in the negotiations to make a difference.

But there's a paradox. We're not only individuals; we're also consumers, citizens, taxpayers, voters, and—if things get bad enough—protestors and sometimes even angry mobs. Only in the aggregate do we have power, and the more we organize, the more power we have.

Even an airline president, while making his way through airport security, has no power to negotiate the level of security he'll receive and the trade-offs he's willing to make. In an airport and on an airplane, we're all nothing more than passengers: an asset to be protected from a potential attacker. It's only outside the system that each of us has power: sometimes as an asset owner, but more often as another player. And it is outside the system that we will do our best negotiating.

• • • •

In his writings, Stanford law professor Larry Lessig has defined four "environmental" constraints on behavior. These same four constraints largely determine the setting in which you make decisions about security and, therefore, your negotiating points, as well.

It is possible to change security by modifying the environment, which in turn changes the trade-offs that the various players make.

For example, a company wants to maximize its profits. If something changed about security—the price went down, more people were willing to pay money to buy it, and so on—the company might change its behavior. Its agenda remains the same, but the constraints on achieving that agenda change. This kind of thing affects your security only indirectly, but significantly. Changing the environment that players use to make trade-offs isn't easy, but we as a society do it all the time. Which method is most effective in any situation depends mostly on non-security aspects of the situation.

1. *Law*. Governments can change the relative power, and therefore bargaining position, of the different players by making certain behaviors mandatory and others illegal, by taxing certain things and subsidizing others, by regulating different behaviors, and by establishing liability laws. Certain border security countermeasures are mandatory; certain home security countermeasures are illegal. Privacy laws require companies to protect personal information. Governments can require banks to accept losses for certain security failures or to cap damage awards for other security failures. Laws also affect security, increasing risks to attackers. Enforcement matters, too. If the police force in your jurisdiction is ineffectual against organized crime, or if the laws in your country allow the government to arrest you with no provocation, there's not much you can do even if the laws are on your side.

2. *Market forces*. Market forces affect security by encouraging or discouraging manufacturers to provide security in their products. Tamper-evident drug packaging was largely a result of marketplace demand for increased security; the lack of security in home computers is also a result of marketplace demand for ease of use. (Note that in both cases customers are making a trade-off.) On the other hand, if enough people want a countermeasure, more choices will become available: nearly unpickable door locks, steel-lined apartment doors, and factory-installed car alarms are readily available because there is a demand for them. There are several problems with this solution, a primary one being that most people are uninformed consumers of security and may be enticed more by the feeling of security than by real security.

3. *Technology*, or what Lessig calls "architecture." Technology affects security in a myriad of ways, discussed in Chapter 7. It changes

security trade-offs by making new countermeasures, and sometimes attack tools, possible. And as new things become possible,
and as they become cheaper, players have security options they
might otherwise not have had. Mounting surveillance cameras on
every street corner simply wasn't an option when cameras were
unreliable and expensive. Now that they can be installed cheaply
and connected using a wireless network, the trade-offs are different. Home burglar alarms experienced a huge jump in sales when
cheap wireless systems became available. Computers and networked databases changed the economics of privacy invasions.

4. *Societal norms.* Societal norms affect security by affecting behav-
    ior—both attack behavior and defensive behavior. In Japan, it's
    perfectly secure to leave your bicycle unlocked outside of your
    home; the country has a strong social network of trust and very
    little theft. In some U.S. cities, the bicycle is sure to be stolen
    within the day. In China, you can openly buy a cheap DVD of any
    current movie, or a copy of any popular college textbook, because
    what many countries call pirating of intellectual property just isn't
    perceived as wrong. Changing societal norms is slow, but it has a
    drastic effect on security. Crime is reduced in areas where property
    rights are better respected. To take an even more extreme example,
    terrorists often come from the fringes of society that preach the
    moral virtue of terrorism. Raising the standard of living is an obvi-
    ous example of a changing societal norm's effect; nothing increases
    security like gentrification. Democracy and property ownership are
    two more examples. In countries where crime is endemic, changes
    like these make an enormous difference in overall security.

All four mechanisms can affect a complex security system. For
example, suppose a corporation wants its employees to clean their
desks of papers and lock them up every night. It can provide technology to make the task easier: combination locks on filing cabinets so
there's no key to lose, or lockable roll-top desks that simply cover over
everything. It can use law: issue a corporate policy and fire anyone
who disobeys. It can change the culture and make employees *want* to
lock their papers up. How effective any of the mechanisms are
depends a lot on the details. Corporations regularly issue "clean desk"
policies, and they are routinely ignored by employees because the culture doesn't accept the policy and there's no enforcement. On the

other hand, when I worked for the U.S. Department of Defense we *always* locked up everything every night: the law was enforced by security guards, and the culture reinforced that it was important.

Cell phones do not provide any real voice privacy. If the law required cell phone companies to implement privacy features, they would. If consumers refused to buy cellphones without privacy features, companies would provide those features. Technologies already exist that can provide good voice privacy in cellphones. But the societal norm is that people generally don't eavesdrop on cell phone conversations. If you want to increase cell phone voice privacy, your best bets for modifying the environment are either law or consumer demand. You could eavesdrop on a lot of cell phone conversations and publish the results, in an attempt to change societal norms, but that seems dicey.

Airline security is no different. Government regulations control airline security. Regulations require airlines to have certain security countermeasures in place. Airline security is identical for all airlines operating in a specific country; you can't choose an airline, or even a flight, that offers more security than the others. No airline advertises "More Secure Airways: Twice as Many Background Checks as the Other Leading Airlines" or "Less Secure Airways: We Get You There Anonymously." The airlines don't get to decide what level of security passengers are willing to pay for, and passengers don't get to decide which airline they fly based on their security countermeasures. But these government regulations are strongly influenced by the market-place; everyone knows that if the government banned all carry-on luggage on airplanes, passengers would rebel. Technology also enters into the equation; countermeasures are chosen based partly on how cheap they are to implement system-wide. Luggage screening became more prevalent as machines to do it became cheaper.

The current airline security process isn't perfect. Because the government has removed the responsibility of security from airlines, the airlines have a different agenda. Their goal is not to do the best security job possible, but rather to do the cheapest job that follows the letter of whatever government regulations they are required to follow.

But if you want to change airline security, you have to negotiate. You can vote and lobby, boycott and buy, invent, or work toward social change. You can do all these things. They're the way you can get power in the negotiation, and they're how you can make your agenda heard.

· · · ·

When negotiating for security, keep in mind that the best player to mitigate a risk is the player—the person, business, or government—who is accountable for it. Security follows the money, and a good security system is one where the security requirements are aligned with the financial motivations of the players. I believe that this should be a consideration in any attempt to change the agendas of players by changing the environment: Make the player in charge of mitigating the risk accountable for that risk.

A simple instance of this is to make salespeople liable for shoplifting losses. You can argue whether this is good employment policy—I don't think it is, and it's a trade-off that the employer has to make—but it certainly makes the people who can best defend against shoplifting interested in defending. As long as the salespeople aren't liable for losses, preventing shoplifting is just another one of their jobs.

Here's another example. Store owners want their salespeople to ring up a sale and provide a receipt, because that practice also generates an internal register receipt and makes it harder for salespeople to steal from the register: It produces an accurate audit trail. Honest salespeople don't care one way or another, and in stores where returns are not common—such as fast-food restaurants—neither do the customers. A common security practice is to put a sign on the register that says: "Your purchase free if I fail to give a receipt." What that sign does is give the customer an interest in paying attention to whether or not she gets a receipt and immediately reporting an employee who doesn't give her one (by demanding her purchase free). It enlists her as a security agent to defend against employee theft. The customer has the capability to perform this security function, and the sign gives her the incentive.

There are many ways to improve the security of personal data in company databases, but the most effective ones force the companies themselves to improve their own security. If, for example, misuse was illegal and companies were liable for the results of data theft, companies that knew they would be fined for misusing customer data would take pains to prevent it. It's simple economics.

Unfortunately, not all security problems can be treated this way. No matter how much it would improve security, airport security screeners can't be made personally liable for airport security lapses, nor does it make sense to arrest them if they make mistakes. What Congress did in the months after 9/11 was the next best thing: They made them federal employees. Before then, private companies working for the airlines employed the screeners. Security was their job, but their employer's interest was primarily to move passengers through as quickly and cheaply as possible. The results were pretty much what you'd expect: poorly paid, poorly trained screeners who did a much better job of looking as if they were effective than actually being effective. Once the security screeners became government employees, things immediately improved. Now the screeners are working for security, not for the airlines. They are more professional. They are better paid. It's still a tough security job for a whole lot of technical reasons—and you can argue about the relative efficacy of making screeners government employees versus commercial contractors to the government—but at least the interests of the security screeners are more in line with the objective of providing better security.

. . . .

One of the ways governments change player trade-offs is through liability. Legal liability is created through certain laws that enforce responsibility and accountability. The courts can enforce liabilities and impose financial penalties for negligence. If there are no real consequences for having bad security, then there's no incentive to improve it. (Remember ATM fraud in the UK? Because the banks were not liable for fraud—consumers were—the banks had no incentive to increase security and fix the problem.) If we expect companies to spend significant resources on security—especially the security of their customers—the security vendors must be liable for security mishaps. If we expect product vendors to market secure products, they must be liable for security vulnerabilities in their products. Liability is an essential component of an environment that fosters security. It increases the direct cost of having bad security and therefore modifies the interests of players.

Liability has its own problems. Sometimes the liability penalties are far greater than what makes sense, and players implement more security than they need because they fear the liability awards. Recall the point I made earlier, that many organizations canceled outdoor events in the DC area while the snipers were still at large. Because the fear of lawsuits was so great, the organizations were forced to implement more security than would otherwise have made sense.

Insurance is a time-honored way to transfer liabilities and therefore manage risk, using the marketplace. The insurance industry pools risk. From the perspective of the person (or company) buying the insurance, it turns a variable-cost risk into a fixed expense. Fixed expenses are much easier on budgets, which is why companies like insurance so much. Insurance may cost an organization more money than paying for the effects of actual losses, but it's worth it to pay for predictability.

In general, the risks worth insuring against have particular characteristics: Risks that are both infrequent and not very serious are not worth insuring against. Risks that are both frequent and serious are probably out of your control. If the river in your town floods every year, you're either going to have to move or wait for the Army Corps of Engineers to come fix the problem. It's going to be impossible or prohibitively expensive to get insurance.

Risks that are frequent and not very serious should be fixed. If your grocery store is experiencing petty shoplifting multiple times a week, every week, you need to install security countermeasures to reduce the amount of shoplifting. You can buy insurance against the loss, but it will cost more than the loss itself. And finally, risks that are infrequent and serious—fire, rare floods, meteor impact, terrorist attack, massive insider fraud—should be insured against. Because they're so rare, that's the most cost-effective way to manage those risks. Most serious security risks fall into this last category.

One of the interesting effects is the rise of the insurance industry as a facilitator of increased security. In many areas of our society, the insurance industry drives security. A company doesn't buy security equipment for its warehouse—strong locks, window bars, or an alarm system—just because it wants to feel safe. It buys that security because then its insurance rates go down. Once insurance companies have enough data, they start charging different premiums for different levels of security. A company will manage its security risk in the most cost-effective way possible; that's simply good business.

· · · ·

As a passive member of society, you largely have to accept the security that is imposed on you by your proxies and by other players. If you want to make your own security trade-offs, you have to work with the agendas of the various players and negotiate as much as you can. This is the only way you can make real changes. If, for example, government getting into the business of massive data-collection on its citizens troubles you, you have to tell your elected officials that you won't vote for them if they support such measures. Or better yet, organize other voters. Or organize as consumers; there's much more power in the organized action of many. If you choose to remain passive, all you can do is complain. If you take an active role in your own security, you have more options that affect your security, and this usually translates into having more options regarding your security.

Recall the cell phone privacy example. You can't buy a cell phone with voice privacy, because the cell phone companies don't have one to sell you. You may have an interest in voice privacy, but the cell phone companies provide only security theater. If enough people refuse to buy products without voice privacy, the phone companies would listen. If the government passed a law mandating real voice privacy, the phone companies would obey. There are no technical issues involved with providing voice privacy, and as soon as the companies have a compelling business interest—such as a demonstrable market willing to pay extra for it—they will do it.

Remember Tylenol and the tamper-resistant packaging? It was never in the pharmaceutical manufacturers' interests to provide tamper-resistant packaging until the marketplace demanded it. Then they did it.

If you are interested in finding a security solution that more closely matches your agenda and your own perceptions of your risks, you need to learn how to negotiate. You need to align the interests of players with each other. You need to align the security interests of players with their capabilities. When you peel away the veneer of feelings about security, real security is all about money. But make sure you understand the financial motivations correctly, because it isn't always obvious how the money flows.

Chapter 17

Security Demystified

Security is more than important; it's an essential and inevitable part of who we are. Because it can never be absolute and static and rigid, it's helpful to think of security as a game—but one that never ends, and one with the most serious consequences. We have to be resourceful, agile, alert players. We have to think imaginatively about our opponents. And we have to move beyond fear and realize that we live in a world in which risk is inherent and failures are inevitable. Thinking sensibly about security requires that we develop a rational sense of the numbers underlying risks, a healthy skepticism about expertise and secrecy, and a realization that a good deal of security is peddled and imposed and embraced for non-security reasons.

Security is a tax on the honest.

If it weren't for attackers, our lives would be a whole lot easier. In a world where everyone was completely honorable and law-abiding all of the time, everything we bought and did would be cheaper. We wouldn't have to pay for door locks, police departments, or militaries. There would be no security countermeasures, because people would never consider going where they were not allowed to go or doing what they were not allowed to do. Fraud would not be a problem, because no one would commit fraud. Nor would anyone commit burglary, murder, or terrorism. We wouldn't have to modify our behavior based on security risks, because there would be none.

But that's not the world we live in. Security permeates everything we do and supports our society in innumerable ways. It's there when we wake up in the morning, when we eat our meals, when we're at work, and when we're with our families. It's embedded in our wallets and the global financial network, in the doors of our homes and the border crossings of our countries, in our conversations and the publications we read. We constantly make security trade-offs, whether we're conscious of them or not: large and small, personal and social. Many more security trade-offs are imposed on us from outside: by governments, by the marketplace, by technology, and by social norms. Security is a part of our world, just as it is part of the world of every other living thing. It has always been a part, and it always will be.

By and large, it's been a successful part. Again and again I've read pundits who maintain that the world since 9/11 has changed utterly—that nothing is the same, least of all security. Of course our sense of threat changed, as did some of our security expectations. But somehow this new feeling of vulnerability morphed into an almost panic-stricken need to change everything. This need is simplistic, too easily accepted, and too ready a justification for all kinds of nonsense and mischief.

The reality is that on 10 September 2001, we had security systems in place that were extensive, complex, effective, and flawed—but not completely flawed and not irremediably flawed. Much of our security worked just fine. FAA procedures cleared the skies over the U.S. within two and a half hours. New York police officers were on hand right away to minimize the ancillary security issues—lootings and other crimes—that often accompany large-scale disasters. Using established procedures and systems, without any new investigative powers, the FBI quickly traced the identities of the terrorists and painted a detailed picture of their actions. Other security agencies were instrumental in arresting terrorists worldwide, freezing their financial assets, and denying them the ability to operate freely. Even the systems that failed on 9/11—national intelligence and airline security—were still largely successful. Yes, the attacks were not prevented. Yes, this was a security failure. No, it was not an utter and dismal failure of the security systems.

In every aspect of our lives, we have a plethora of security systems in place and the experience and expertise that goes with it. None of these systems is perfect, but none of them is unsalvageable, either. The challenge is to figure out what to keep, what to alter, what to toss, and what to build from scratch. It is not merely idle pride and foolishness to assume you have to reinvent the wheel just because of a single failure; it is also a terrible mistake—as serious a mistake as blindly thinking that the status quo is fine.

The status quo is never fine, because security is never done. It's constantly in flux, and the best security systems are the ones that evolved and continually evolve. It's like the old saying: "You can't win. You can't break even. You can't get out of the game."

And "game" is a good metaphor for security. As in a real game, there are two sides—attackers and defenders—and the other side is a live, reacting, intelligent opponent. As in a game, both sides can devise strategies to defeat their opponents. As in a game, there are surprises when one side invents a new strategy and the other side has to react.

As in a game, there are moves and countermoves, both within a single attack as defenders try to defeat it, and longer-term as attackers develop and implement new attacks and defenders develop and field new defenses. As in a game, rigidity is not an asset.

Of course, security is not a lighthearted game like Monopoly or miniature golf. Money, lives, and even our way of life can be at stake. If security is a game, then it is a game for mortal stakes. But this is not to say that security should be conducted in a fearful or fatalistic manner; sensible security is never rooted in fear.

There are many ways in which security is not like a game. Unlike most games, strategies are constantly changing. Defenders might be ready for a certain kind of attack, only to face something else instead: An attacker with a different goal. A better-funded attacker. An attacker who is willing to kill. Defenders may face a variety of different attackers at the same time. Actually, security is worse than a game with constantly changing strategies; it's a game with constantly changing rules.

Unlike a game, security has no beginning and no ending, no winner and no loser. It's a mistake to think that security is an achievable, fixed, static state. It's simply not possible to get security right once and for all, because the meaning of "right" is constantly changing. This situation has been with us since we humans started forming communities, and it's likely to be with us as long as we continue to be human.

There are three basic categories of information: things we know, things we don't know, and things we don't even know we don't know. The last category is what makes security so difficult. We might know the inner workings of an organized crime family and its goals, and be in a good position to evaluate trade-offs and implement sensible countermeasures. We might not know Al Qaeda's next moves, which makes it much harder to defend against them, but we have a reasonable expectation that it's plotting something. But the hardest security problems of all result from the things we don't even know we don't know: new technologies that can be used in surprising ways, new twists on old ideas, and new attacks we've never thought of. And even completely new attackers—enemies we don't yet know are enemies. I consider these to be failures of imagination—failing to think like the attackers. If security is a game, it's against the smartest of opponents. And the only way to face the attackers of the future is to be alert, agile, and responsive.

Security can never be absolute. How many times have you heard the recent canard, "No one will ever be able to hijack an airplane again

unless he is able to kill every passenger on board first"? Nice sentiment, but it's just not true. I can imagine lots of ways airplane hijackings can still occur: a plane that's empty enough that the hijackers outnumber the passengers, a hijacker who succeeds in convincing the
passengers that he's not suicidal or a terrorist (carrying a baby would
go a long way toward calming the passengers), a hijacker who succeeds
in taking over a bulletproof cockpit (turning a security countermeasure
into a vulnerability), or a hijacker who convinces everyone that she's a
sky marshal. The point here is not to speculate on human psychology,
but again to demonstrate that all countermeasures are capable of failing. Words like "always" and "never," when used to describe security
solutions, are major contributors to bad security decisions. The people
on United Flight 93 were heroes; I would like to think that the next
group of people in a similar situation would also be heroes, but I don't
know how to guarantee that it will be true.

Sensible security is commensurate with the risks. A velvet rope
and a bouncer is enough security to protect the VIP room at a dance
club. A few clerks and societal honesty is enough to defend a grocery
store against shoplifting. Door and window locks are enough security
for many homes. Protecting a head of state requires extensive security
systems. A movie star, more concerned about fans than assassins,
requires a less extensive system.

The idea that there is some Holy Grail of security is naïve; it prevents progress of any sort and destroys freedom. Those who pursue the
ideal of ultimate security do so to the detriment of everything else, and
the trade-offs they make can border on the absurd. Risk is inherent in
life, and the price of freedom is the possibility of crime.

····

As both individuals and a society, we can make choices about our
security. We can choose more security or less security. We can choose
greater impositions on our lives and freedoms, or fewer impositions.
We can choose the types of risks and security solutions we're willing to
tolerate and decide that others are unacceptable.

As individuals, we can decide to buy a home alarm system to make
ourselves more secure, or we can save the money because we don't consider the added security to be worth it. We can decide not to travel
because we fear terrorism, or we can decide to see the world because the
world is wonderful. We can fear strangers because they might be attackers, or we can talk to strangers because they might become friends.

Number of Deaths, Worldwide, from Different Diseases

HIV/AIDS	3,000,000 annually	Estimates vary widely. Another 42,000,000 people are infected.
Tuberculosis	1,900,000 annually	Drug-resistant strains of the disease have emerged.
Malaria	1,000,000 annually	There are higher estimates.
Influenza	250,000 annually	Annual epidemics are thought to result in 3 to 5 million cases of severe illness and 250,000 to 500,000 deaths, even though a moderately effective vaccine is available.

Note: Estimates are based on data from developed countries.

Odds of Different Things Happening to a Person, in a Year, in the U.S.

1 in 85	Being injured in a car accident
1 in 230	Having a car stolen
1 in 340	Being abducted by a family member (children under 18)
1 in 670	Being robbed
1 in 1,200	Being abducted by a nonfamily member (children under 18)
1 in 1,600	Being a victim of forcible rape (females, all ages)
1 in 17,860	Being murdered
1 in 240,000	Being struck by lightning
1 in 56,624,800	Dying of anthrax

Note: These numbers are culled from a variety of sources, for years ranging from 1999 to 2002, and are often rounded.

As a democratic society, we can decide that constraints on law enforcement power are good for society, and that living in a free and open society is more secure than living in a totalitarian police state. We can decide that racial profiling is not the kind of thing we want to do in our society, or that the dangers are worth the societal inequities caused by profiling. We can decide not to torture because it is immoral, or we can decide that the information it gives us—flawed as it is—outweighs its immorality and the long-term loss of security due to the inevitable abuses. We can decide to ban certain devices or technologies because we fear their effects in the hands of attackers, or we can make those same devices and technologies available to all because we know how beneficial they are.

Real security requires people to understand and appreciate attackers, attacks, countermeasures, rules, procedures, systems, and the agendas of all players involved. It involves a sense of the numbers: what the risks are, how well the countermeasures mitigate them, and what the trade-offs are and what they're worth. It involves thinking sensibly about security, not thinking fearfully.

And sometimes we get it wrong. Sometimes we underestimate the risk and implement less security than we need. Sometimes we overestimate the risk and pay more in trade-offs than we have to. When this happens, we reevaluate and make changes. We ratchet security either up or down to correspond to the real risks.

This process is no different from any other decision we make as individuals and as a society. We're constantly trading off one thing against another: choosing one career over another, not getting a flu shot one year, sending our children to this school and not that school. As a society, we do the same things on a larger scale, deciding whether to fund different social programs and whether to encourage different behaviors with laws, regulations, or tax breaks. When we make mistakes, either as individuals or as a society, we do our best to fix them.

Sometimes security failures occur even when we get the evaluation process and the security level right. We can take more precautions than reasonable, spend more money than prudent, suffer more trade-offs than normal, and still be attacked. Finding blame or fault is a perfectly human reaction, but it's important to accept that sometimes failures simply happen. Not all grievances can be redressed. Not all wrongs can be righted. Not all things can be fixed. This fact can be tremendously serious and heartbreaking. When a sniper attack makes the front page or a rare surgical accident debilitates someone you know, it's natural to demand safeguards so the problem won't happen again. But in a country of 290 million people, even incredibly unlikely events will occur once in a while. That doesn't necessarily mean that the system is flawed, someone is at fault, or retribution is deserved. Too often people imagine all sorts of horrific scenarios and then demand to be protected from them. Lawyers exploit this misconception, as do politicians. Even warning labels on the products we buy (seen on a Domino's Pizza box: "Caution! Contents hot!") imply that somehow we can avoid the risks inherent in life.

Accepting risks is something we do constantly. The only reasonable thing to do is to build security systems that recognize this reality and allow us to go about our lives the best we can. Every time we leave our house, we accept the risk that we might be mugged or attacked. Our countermeasures reduce the risk, but they never reduce it to zero. If we want to leave our house, we have no choice but to accept that residual risk. Ten years ago, I was mugged in Istanbul. I still travel extensively, albeit with a diminished sense of invulnerability. Similarly, as a nation, we can only reduce the risk of terrorism; we cannot eliminate it. We can only reduce the risk of counterfeiting, or credit card fraud, or gang-related crime. Anyone who tries to entice you with promises of absolute security or safety is pandering to your fears. It's important for you to move beyond fear and be skeptical, contentious, provocative, even rancorous. Security is a process, an ongoing game. It's continually evolving. The risks we accept, either by choice or by necessity, are constantly changing. But they are always there.

．．．．

This book often has been negative in tone, and it's hard not to feel discouraged by the sheer multitude of threats and dangers. But the book is about security, and security concerns itself with attacks and losses. It centers on the bad in people. It focuses on mistakes. It aims mainly to stop things, not to make things happen. As a topic, it's a downer.

When lecturing on security, I often pose this question: Civilization has been around for 5,500 years, and murder has been a constant security problem. When will technology be good enough to prevent murder?

The answer is: Never. There's no set of laws, no technologies, no market forces, and no amount of social change that will completely prevent murder. If someone wants to kill you, he very likely will. Period.

Then I ask another question: If that's true, why do we all feel so safe?

It's not because we wear body armor or live in fortresses. Technology is a critical part of security, but it is not our savior. I believe that most of us intuitively know this. We realize that there isn't some magic anti-murder dust we can sprinkle over our bodies to protect us. We know that locks and alarms offer only so much protection. For real safety, we live in nice neighborhoods where people watch out for each other. Or we fund our police departments so that they can maintain a presence in our neighborhood and catch criminals. Or we stay alert and watch for suspicious goings-on, ask "Who is it?" before we open the door, and don't make it obvious when no one is home. In all of these cases, people are the essential element. The best security systems revolve around people.

Intuition is at least as critical to securing our nation as the latest multibillion-dollar technological system. U.S. customs agent Diana Dean's notion of hinkiness cannot be bureaucratized and replaced with hinkiness awareness training, a hinkiness threat index, and hinkiness alerts. Customs agents can be trained, but in the end we're hiring them for their intelligence and intuition—and their ability to develop and sharpen both.

But all security considerations aside, we all go through our lives feeling relatively safe because the overwhelming majority of people are not out to do us harm. We have all sorts of security systems in place defending us against all sorts of threats, but the primary thing keeping our society from complete anarchy is human nature.

If someone is living in fear—whether it's fear of the burglar on your block or the fanatical dictator half a planet away—it's because she doesn't understand how the game of security is played. She longs for a fortress, for a fairy-tale solution that will work forever after. It's a perfectly reasonable longing, and it's because she thinks about security in terms of absolutes or magnifies her level of risk based on her experiences with the media, both news and fiction.

There's a smart way to be scared. It involves moving beyond fear and thinking sensibly about trade-offs. It involves looking beyond the newspaper headlines and getting a feel for the numbers: a feel for the threats and risks, and the efficacy of the countermeasures. It involves making sensible security trade-offs. The smart way to be scared is to be streetwise.

Anyone can understand security. The people who think they know best, and the people who think they ought to, would have you believe that security is like quantum mechanics or brain surgery, and that it should best be left to the experts. These people are wrong; they're trying to co-opt your agenda and increase their own power in negotiating trade-offs. Don't accept claims that you can't understand security. You can. Don't accept claims that the technology is too complicated for you to understand. The details may be complicated, but the ways in which the system works and fails won't be.

Sensible security is something you do every day; you constantly make security trade-offs. Your brain is hard-wired to do it; it's part of being alive. The five-step process, the examples of security failures, the descriptions of attacks and attackers, and the analyses of different countermeasures are my attempts to demystify security and formalize the thinking you already know how to do. And I want to leave you with three final rules:

- *Schneier Risk Demystification:* Numbers matter, and they're not even that difficult to understand. Make sure you understand the threats. Make sure you understand the risks. Make sure you understand the effectiveness of a security countermeasure and all of the trade-offs. Try to think of unintended consequences. Don't accept anyone saying something like: "It would be terrible if this sort of attack ever happens; we need to do everything in our power to prevent it." That's patent nonsense, and what someone living in fear says; you need to move beyond fear and start thinking about sensible trade-offs.

- *Schneier Secrecy Demystification:* Secrecy is anathema to security for three reasons: It's brittle, it causes additional security problems because it conceals abuse, and it prevents you from having the information you need to make sensible security trade-offs. Don't accept anyone telling you that security requires keeping details of a security system secret. I've evaluated hundreds of security systems in my career, and I've learned that if someone doesn't want to disclose the details of a security system, it's usually because he's embarrassed to do so. Secrecy contributes to the "trust us and we'll make the trade-offs for you" mentality that ensures sloppy security systems. Openness demystifies; secrecy obscures.

- *Schneier Agenda Demystification:* People often make security trade-offs for non-security reasons. Therefore, it is critical to understand the agenda of players involved in a security negotiation. Don't blindly accept that "It's for security purposes" when someone tries to justify a questionable countermeasure; far too often players attempt to disguise their actions as security-related. Understand how emotions lead people to make seemingly irrational security trade-offs. Agendas are natural, and they're pervasive—be aware of them.

Fear is the barrier between ignorance and understanding. It's paralyzing. It makes us fatalistic. It makes us do dumb things. Moving beyond fear means freeing up our intelligence, our practical common sense, and our imagination. In terms of understanding and implementing sensible security, moving beyond fear means making trade-offs openly, intelligently, and honestly. Security is a state of mind, but a mind focused on problem-solving and problem-anticipating and problem-imagining. Security is flexible. Fear is also a state of mind, but it's brittle. It results in paranoia, paralysis, and bad security trade-offs.

An unfortunate truth is that because we are human, we are neither perfect nor perfectible. But we can improve. And historically, both attackers and defenders have improved. Historically, the security advantage oscillates between attacker and defender, with defenders slowly and continually making progress. The one thing defenders have going for them is that there are far more of them than there are attackers.

And this is what makes society work out in the end. Attacks are the exception. Because we live in a civilized society, we are primarily moral and ethical. And because of that, most of us are generally safe and secure. It's important to keep this in mind. As horrible as the newspaper headlines might be this week, remember that the one or two or twenty victims are the exceptions, and the other 291,858,619 people (projected U.S. population on 5 September 2003) did not suffer the same ordeal. I don't mean to trivialize the suffering of the victims, but to put it into perspective. As scary as the threats sound, accidents are a much greater risk. The state of being secure is the normal one.

Can you make your home absolutely safe? No; that's the bad news. Can you do things to dramatically reduce the risk of having bad things happen to you and your family? Yes; that's the good news. But the things you can do generally have more to do with protecting against accidents than protecting against attackers.

When people think of keeping their families safe, they mostly think of protecting them against intruders. They think of preventing the kinds of horrible things that make the front pages of newspapers. But those aren't the real risks. Horrifying as those crimes are, they are not what kill most kids. The leading causes of death to children are automobile accidents, gun injuries, drowning, and suicide. The likelihood of each can be reduced greatly, and by means that everyone already knows. Buckle up your kids. Lock up your guns. Pay for swimming lessons and teach your kids to swim. Pay attention to your children's state of mind. Don't tolerate abuse.

Still scared of a crazed stalker breaking into your house and kidnapping your children? That's okay, but make sure you know that the fear is in your mind and not based on the real risks. The best way to protect against intruders in your home is to turn on the bathroom light. Don't believe me, believe the world's smartest ex-burglar, Malcolm X: "I can give you a very good tip if you want to keep burglars out of your house. A light on for the burglar to see is the very best means of protection. One of the ideal things is to leave a bathroom light on all night. The bathroom is one place where somebody could be, for any length of time, at any time of the night, and he would be likely to hear the slightest strange sound. The burglar, knowing this, won't try to enter. It's also the cheapest possible protection. The kilowatts are a lot cheaper than your valuables."

Or get a dog....

Now *that's* a security trade-off.

Author's Note

Security is an ongoing process; it's never done. Because of the realities of publishing, this book represents a snapshot of my thinking in spring 2003. By the time you read this, things will have changed. There might be new political realities. There might be new security ideas, new attacks, new analyses. The best security systems continually evolve. My thinking about security is no different.

Every month I write and publish an e-mail newsletter called *Crypto-Gram*. It's a compilation of news and essays on security, much along the same vein as this book. In fact, many of the ideas and concepts in this book began as *Crypto-Gram* essays. Any new analyses of mine will appear in *Crypto-Gram* before they appear anywhere else.

If you're interested in my continuing commentary on security and security-related topics, please subscribe to *Crypto-Gram*. I already have 80,000 readers, and I welcome more.

To subscribe to *Crypto-Gram*, visit:

> http://www.schneier.com/crypto-gram.html.

Or send a blank message to:

> crypto-gram-subscribe@chaparraltree.com.

Back issues of *Crypto-Gram*, as well as my other writings on security, are available at:

> http://www.schneier.com

Thank you for reading this far. I appreciate your willingness to accept me as a teacher and a tour guide, and I hope you have learned much about security as a result of our time together. If you have any questions, comments, complaints, or criticisms, please feel free to send me e-mail at schneier@counterpane.com. I don't promise that I will have the time to answer every e-mail, but I do promise to read them.

And remember to use the five steps. Practice analyzing security today. Look at the security countermeasures in place to prevent theft at a gas station. Examine the security countermeasures that stop someone from stealing free power from the electric company. Whether it's a velvet rope and a bouncer in a nightclub, or a metal detector and an armed guard at an airport, you can understand how security works and fails.

Acknowledgments

Writing this book was a surprisingly social task, and many people had a hand in its creation. I would like to thank Charles Mann, whose article about me in *The Atlantic Monthly* (November 2002) precipitated this book, and whose continuing conversation and counsel helped me determine the tempo and timbre of the book. I would also like to thank my agent Kris Dahl and my first editor Jonathan Segal. The people at Copernicus have been great to work with, and this book would have never seen the light of day without the hard work and dedication of my editor there, Paul Farrell. No less important was the work of my assistant and personal editor, Beth Friedman. Research, editing my writing, editing other people's edits, proofreading, beating my word processor into shape: I can't even begin to list all the things she was responsible for. Raphael Carter and Vicki Rosenzweig also deserve thanks for their research assistance, as does everyone at Counterpane Internet Security Inc. for giving me the freedom to write this book. And finally I would like to thank my wife, Karen Cooper, for putting up with the roller coaster ride this book has been.

A veritable battalion of people read this book in various stages of completion. In particular, I would like to thank Michael Angelo, Phil Agre, Steven Bass, David Brin, Steven Brust, Jon Callas, Tracey Callison, Raphael Carter, Vint Cerf, Margaret Chatham, Ralph Chatham, Karen Cooper, Tammy Coxen, Lori Curtis, Walt Curtis, Dorothy Denning, Steve Diamond, Cory Doctorow, John Donat, Jeff Duntemann, David Dyer-Bennet, Carl Ellison, Rik Farrow, Charles Faulkner, Amy Forsyth, Jerry Forsyth, Dan Geer, Barry Gehm, Jennifer Granick, Richard Graveman, Alis Gray, Gregory Guerin, Anne Harrison, Bill Herdle, Peter Hentges, Bill Higgins, Kelley Mathews Higgins, Julie Holm, Peter Honeyman, Steve Hunt, Jim Huttner, Ericka Johnson, Lance Johnson, Angel Johnston, Chris Johnston, James Johnston, James Jorasch, Mich Kabay, Jonathan Katz, Michael Katz, John Kelsey, Peg Kerr, Tadayoshi Kohno, Scott Laiken, Denise Leigh, Stephen Leigh, Elias Levy, Steven Levy, Rachael Lininger, Charles Mann, Elise Matthesen, Anna Meade, Jay Miller, Doug Morgenstern, Tim Mullen, Anne Murphy, Bill Murphy, Peter Neumann, Patrick Nielsen Hayden, Al Nofi, Andrew Odlyzko, Pierre Omidyar, Charles Perrow, David Perry, John Pescatore, P.M. Picucci, Michael Pins, Sean Rooney, Greg Rose, Vicki Rosenzweig, Martin Schafer,

Rebecca Schneier, Lee Short, Jenna Sindle, Stephan Somogyi, Holt Sorenson, Ken Sprague, Mike Stay, Jennifer Stevenson, Samantha Star Straf, Richard Thieme, John Thompson, David Wagner, Jay Walker, Toby Weir-Jones, Brad Westervelt, Doug Whiting, Moti Yung, and Elizabeth Zwicky, all of whom read and commented on the book in one—or more!—of its many draft stages. These people did a lot to make the book complete, accurate, and interesting. Any remaining omissions, lingering errors, or residual prolixity are solely my own fault.

Index

Entries beginning with 9/11 are listed under "n" for "nine"

A

D

data
  aggregation of, 98, 99, 253–54
  analysis of, 161–63
  authentication of, 196–98
  information distinct from,
    161, 162, 163
  protection via replication, 217
  theft liability, 267
data mining, 98–99, 253–54
databases
  abuses of trust, 138
  attacks facilitated by, 77–78,
    99
  authorization status incorpo-
    rated in, 193
  CAPPS, 165
  in counterterrorism, 243, 247
  of credit card numbers, 84
  DMV, 77–78, 193
  for forensic auditing, 158
  identifiers in, 185–86
  identity theft facilitated by,
    75, 243
  liability for, 267
  misuse of, 205
  national ID cards, 204, 205
  spheres of trust, 140–41
  TIA program, 253–54
Dean, Diana, 134, 136, 278
deaths
  causes of, 29, 30, 238, 275,
    280–81
  from disease, 27, 30, 238, 275
  due to terrorist attacks, 3, 27,
    227, 239–40
deception, 213–15
decoys, 213
default to insecure system behav-
    ior, 128–29
defenders
  concept of, 12
  majority status of, 280
  power imbalance with attack-
    ers, 87, 89, 101
  reactive status, 104
  technology and, 101
defense in depth
  for counterterrorism, 252
  division of trust as, 141
  in military strategy, 107

in natural world, 121
offense distinct from, 173, 175
in product packaging, 116
purpose of, 103, 105
to secure trusted people, 141
defense strategies
  choke points, 103, 109–10
  compartmentalization, 103,
    105, 107–9, 116, 140–41
  defense in depth, 103, 105,
    107, 116, 121, 141, 252
  determination of weakest
    link, 104
  overengineering, 107
  'position of the interior', 104
  see also countermeasures;
    responses
democracies, 11, 244, 246
demystification rules, 279–80
Denny's, 229–30
department stores *see* retail stores
detection countermeasures
  active failures, 153
  advance detection, 160–63, 243
  for anonymous transactions,
    159
  for computer files, 160
  in counterterrorism, 242, 243
  passive failures, 153
  probing of, 154–56
  random detection systems,
    150–53, 156
  rarity-based failures, 56–57,
    137, 153–54, 167, 228
  retrospective, 156–59, 160, 216
  in security triad, 147, 148–50
  technology in, 154
  types of, 150–52
  visibility of, 151, 219
deterrence
  in counterterrorism, 243
  efficacy of, 176–78
  positive bag matching as, 220
  reduced media coverage as, 70
  as a response, 167, 168
  retrospective detection as, 12
DHS *see* U.S. Department of
    Homeland Security
    (DHS)
diamonds, 119, 142, 211–12
disaster recovery plans, 227, 230

disease, deaths from, 27, 30, 238,
    275
distaste, as countermeasure,
    220–21
diversification, 124, 125
diversions, tactical, 170, 211–13
DMV (Department of Motor
    Vehicles) database, 77–78,
    193
double-entry bookkeeping,
    156–57, 160
driver's licenses, 114, 192, 193,
    194, 196
Driving While Black (DWB),
    135
drug dealers, 63, 77, 79, 108, 221,
    247
dynamism
  in natural world, 122
  of people, 133, 145–46
  in security systems, 119, 121,
    122, 123–24, 125

E

ear shape, in biometrics, 187
East Germany *see* German
    Democratic Republic
    (GDR)
economic targets, 235
education, as countermeasure,
    167, 168, 178, 243
Einstein, Albert, 112
El Al Israel Airlines, 115, 151–52
elevator systems, 48
Ellsberg, Daniel, 63
emergent properties, 49, 50, 112,
    219
emotional attacks, 60, 64–65, 176
employees, attacks by, 62–63
engineering principles, 111, 112
engineering systems, 50–51
Enigma machine, 58, 95
enrollment systems, 192–93
entrapment, 64
envelopes, 216
espionage, 58, 65–66, 95, 174
espionage agents, 62, 63, 142–43,
    214
ethics, 178, 263
evaluation process *see* risk analysis
    and evaluation

Printed in Great Britain
by Amazon